SUFFOLK RIVER BRIDGES

1700 - 2017

Carol Twinch

By the same author

Women on the Land: Their Story During Two World Wars
In Search of St Walstan: East Anglia's Enduring Legend
Tithe War 1918-1939: The Countryside in Revolt
Great Suffolk Stories
Saint with the Silver Shoes: The Continuing Search for St Walstan
Bury St Edmunds: A History and Celebration
Walk Around Historic Bury St Edmunds
Ipswich: Street by Street
Little Book of Suffolk
The History of Ipswich
Essential Essex
Walks Through History: Ipswich
The Norwich Book of Days
Saint Walstan: The Third Search
Early Bridges of Suffolk

Plain and Simple Egg Production
So You Want to Keep Sheep
Poultry: A Guide to Management

First Edition 2017

Published and Printed by
Leiston Press Ltd
Masterlord Industrial Estate
Leiston
Suffolk
IP16 4JD
Telephone Number: 01728 833003
Web: www.leistonpress.com

ISBN: 978-1-911311-22-5

CONTENTS

'Friars Bridge, Ipswich - this stream joins the Gipping' by H Davy (1837)

'Bungay, Suffolk' an engraving by J Greig
from a drawing by J S Cotman for the Excursions Through Suffolk (1818)

Front Cover: the bridge and extended footbridge at Sapiston where
excavations in the 1940s identified Neolithic and Bronze Age occupation.

INTRODUCTION

Like Suffolk's parish churches each one of the county's river bridges is unique. That isn't to say that there are not hundreds of small, work-a-day bridges similar to one another, but all have their own characteristics and history. When first embarking on bridge research I met a friendly and helpful man from the Environment Agency at one of the gauging stations. On hearing my mission he raised his eyebrows and declared 'you've got your work cut out!' and it turned out he was quite right. Once you start looking, there are bridges everywhere over watercourses of every conceivable size from little more than a ditch to the River Orwell. Then there are the legion of bridges which are relics of 19th century navigation schemes, as well as modern bridges at gauging stations and completely new highways as at Tattingstone.

It has not been possible, of course, to write about every bridge in Suffolk, or mention even half of them, although some bridge history has already featured in *Early Bridges of Suffolk*. As well as those on regular carriageways, there is a profusion of small brick bridges in unexpected places. Along the lane leading from Hoxne school to the top road coming from Stradbroke is such a one. Just outside Framlingham, on the B116 to Wickham Market, a beautiful little hump-backed, single-arched bridge marks the route of the old road: it still carries a now illegible keystone and its parapets and capping bear the scars and scrapes of heavy traffic sustained, no doubt, in the final years when it carried the main road before the new bridge was built.

Often, a seemingly modern and unremarkable bridge at road level hides an attractive one or two-arch structure beneath. At Little Bealings, for example, the apparently mundane crossing over the River Fynn masks an enchanting white-bricked bridge reminiscent of another age. Some have an unexpected curvature, such as that at Ubbeston, showing a commendable degree of engineering that is unsuspected from the modern parapets lining the roadway. The design at Heveningham is based on that at neighbouring Ubbeston and has the same graceful curve.

Thankfully, over the years since increased traffic has necessitated strengthening existing bridges councils have often merely placed steel girders across the top and left the basic structure intact. Yet, alas, many small bridges were wiped from the landscape, existing only in old photographs or drawings. An engraving of 1780 by Richard Godfrey shows the East Gate Bridge in Bury St Edmunds. It appeared in the *Antiquarian Repertory,* a miscellany 'intended to preserve and illustrate remains of Old Times'. The East Gate itself had been taken down in the 1760s but the five-arched bridge, not far from the extant Abbots Bridge and contemporary with it, was taken down in 1840. There were, no doubt, many sad to see it dismantled though perhaps others saw it as progress. Its replacement, the Steggles Bridge, faired somewhat better when, in 1986, the bridge was threatened with demolition after it was blamed for causing floods in the town. Local residents, however, objected and won the day.

A few years earlier in 1959, the imposing 19th century Seven Arches Bridge over the River Orwell in Ipswich was demolished. It stood more or less where the Hadleigh and Yarmouth roads meet and was one of those bridges which carried an Ordnance Survey Pivot Bench Mark, sited on top of its southern battlement. (More about the county's bench, or cut, marks can be found on the Bench Mark Database.)

Sometimes, though, even where the bridge is modern, if it is an ancient crossing the banks and river bed often contain evidence of earlier structures, for example, at Ballingdon where archaeology has exposed remnants of previous bridges.

During the late 19th and early 20th century, existing parapets were raised with new brickwork, or railings added at road level. White railings are ubiquitous indicators of a bridge, often marking a brook or ditch crossing - like those at Dalham, Laxfield and Grundisburgh - or the multitude of small one and two-arched bridges like the one over the River Box near Great Waldingfield or that over the River Alde at Bruisyard. The tiny little bridge at Dorking Tye (in the parish of Assington) is more like a culvert; unfortunately in 2016 a chunk of the low parapet on the outside of the railings was smashed by a mechanical device, exposing the crimson brickwork.

A significant feature of many traditional bridges is their use of adjacent sites for gauging stations and associated sluices or weirs, looked after by the Environment Agency. In the early 1950s my father carted corn from Hoxne to Billingford Mill for grinding, passing over the little red brick bridge from Suffolk to Norfolk. It is now the location for a large gauging station that monitors the River Waveney.

River crossings evolved from fords to wooden bridges, then to brick and stone until, following the Industrial Revolution, cast iron and then steel and concrete took their place in the evolution of bridge building. Amazingly, several of the county's cast iron bridges are still in use, such as the Scotland Street Bridge at Stoke by Nayland, Sir William Cubitt's three-span bridge at Clare, and the Garrett bridge at Great Thurlow. Although cast iron bridge beams and structures were innovative, it was a relatively short-lived technology. In their day, though, these and others like them were in the forefront of groundbreaking technology. Suffolk was, after all, home to the mighty engineering firms of the Garrett and Ransome families and the genius of William Cubitt and his ilk.

In modern life we can do without quite a lot of things, but bridges are not one of them. Ask the people who have to churn through the centre of Ipswich when the Orwell Bridge is closed; or the people of Tadcaster in North Yorkshire who on 29th December 2015 saw their 18th century bridge crumble as the River Wharfe rose to historic levels. Without the bridge it was a 9-mile detour to get from one side of the river to the other: unsurprisingly there was great rejoicing when the bridge was re-opened in February 2017.

Just as, no doubt, travellers of old complained that the bridges were insufficient for their needs, not wide enough for their pack horses, dangerous or impassable due to weather conditions, so now the traffic build-up causes chaos when normally busy bridges are closed for repairs or high winds forces Highways England to close the Orwell Bridge. There is regular talk about a by-pass to relieve the latter problem as well as the prospective new bridges in Ipswich, outlined in the 2016 Budget. My schooldays were spent at Lowestoft where complaints about the bridge were legendary. After fifty years of lobbying, a new crossing is planned here, too, for some time in the near future. I must have walked over the old bridge many times as our school 'crocodile' progressed from the south of the town to the north and back again, but my memory is more of the trawlers and the smell of the fish market.

The people who witnessed fords being replaced by wooden bridges, and subsequent generations who saw stone and brick take over from wood, approved the upgraded means of getting from A to B. Today not everyone sees the destruction of a small but ancient bridge in favour of iron girders or reinforced concrete as progress, rather the opposite. It is seen as the loss of yet another intricate part of the evolved balance of the Suffolk landscape. Postcards of the 1930s, 40s and 50s show now lost hump backed bridges and what better portrays the rural scene of earlier centuries than an ancient bridge and a ford alongside where cattle and horses were watered?

When it comes to change society's opinions are fickle: in the first decade of the 21st century opinions as to what constitutes progress is mixed and a small, single-arched bridge on a busy road means different things to different people. For instance, who can say that the little bridge at Coddenham is easily negotiated by a large vehicle? Indeed it is not but its removal would be another irreplaceable loss of a tiny part of the beauty and charm of Suffolk. It is all a question of balance and compromise. Whereas in past centuries the rich and powerful had the say, now it is down to individual groups with the ability to lobby authority if a cherished landmark is under threat. Whether or not the array of Suffolk's bridges still left to us survives into the future is down to how much they are valued by the communities where they exist and by those who care about the individual culture and history that they represent.

Many thanks to all the kind and helpful people we met along the way including staff of the Suffolk Library Service, especially Saxmundham Branch; the Environment Agency; Suffolk County Council Highways Department; Long Shop Museum; Ancient House Museum of Thetford Life; Professor Steven Gunn; and my husband Christopher who poured over maps with furrowed brow and eventually did battle with my syntax invariably scribbling 'Who he?' or 'What does this mean?' or, mostly, just a question mark in the margin.

Carol Twinch
Rendham, 2017

CHAPTER 1

When the topographer John Kirby published *The Suffolk Traveller* in 1735 he had spent three years surveying the whole country. He included only nine bridges on his Suffolk maps but, like Robert Reynes before him, names many more throughout his survey as way marks much as he uses inns, both easily identified in the landscape and unlikely to change position. Having visited Southwold he writes:

> Thus much for Southwould. Returning back over the Bridge, at 4 ¾ f[urlong] the right as aforesaid goes over Potters Bridge to Lowestoft. At 5 f[urlong] the left leads over Wolsey-Bridge to Halesworth.

Near Lowestoft itself he says, 'Normanston, now corruptly called Nomans-Town, lies between Mutford Bridge and the Town of Lowestoft'.

Surveying the road from Ipswich to Cattawade Bridge he says:

> From the Market-Cross passing through St Nicholas's Street, at 3 ¾ f[urlong] is Stoke Bridge. At 1 m 6f[urlong] is Bourn Bridge, the left by the Waters side goes to Shotly.

Bourne Bridge is one of the few marked on Kirby's map.

Neither Kirby, nor any of his predecessors, was the first to use bridges as a means of identifying points in the landscape. In the pre-Reformation Dodnash Priory charters, for example, there are several mentions of bridges as landmarks, especially in the matter of administration of legacies of marshes or arable land bestowed on the priory. **Brantham** bridges mentioned include the 'Walne' (1196), 'Cotesbreg' (1297) and 'Constables Bregge' (1363).

In around 1196 a piece of woodland is identified as being 'all his land between the road from the bridge of 'Hwolne' up to the elm-grove to the east and to Dodnash wood' a description not unlike any of the later cartographers and surveyors of the 18th century.

Bridges had traditionally been used as geographic markers in wills, such as that in **Hoxne** when John Elys of Oakley left money in 1375 for the repair of the chapel of St Edmund and 'one rood of pasture lying near the bridge of Hoxne in perpetual arms.'

Kirby makes no comment on the condition of the bridges, their attractiveness or otherwise and only makes mention if he passes over them. On other surveys he often describes the road as being from (say) 'Wilford Bridge to Snape Bridge', again in the style of Robert Reynes. Bridges had gone from being an indispensable and treasured part of local society and landscape to a necessary component in commerce and communication, but one largely taken for granted.

Not that the men of **Ubbeston** and **Heveningham** took their bridge for granted as in the winter of 1793-94 the floods swept away the old wooden bridge cutting off the roads between the two parishes. In June 1794 it was decided that 'a new and substantial bridge of brickwork' should be built. Local Surveyors Robert Baldry and Richard Randall consulted with those in **Huntingfield** who had recently built their own bridge. The design of the Huntingfield Bridge was considered appropriate and an agreement was drawn up, the sum of £67 being proposed to defray the costs. In fact the Ubbeston Bridge has a more graceful curve than that at Huntingfield but the design is similar.

Sir Joshua Vanneck, 3rd Baronet Huntingfield of Heveningham Hall, agreed to contribute £10 towards the project. Sir Joshua's contribution seems somewhat disproportionate given that he had recently inherited estates estimated to be worth £8,000 a year and assets worth between £200,000 and £300,000. In 1792 the *Public Advertiser* stated that, 'Sir Joshua Vanneck, if not certainly the richest commoner in this kingdom, has nearly the largest income. The family estate is £14,000 p.a., the profits of his trading house are sometimes £20,000.'

However, £10 seems to have been the amount he agreed. The two village rectors donated £5 from the glebe lands, and the men from Ubbeston and Heveningham gave varying amounts.

In 1994, on the 200th anniversary of the building of the bridge, Veronica Baker-Smith wrote:

> Careful estimates were drawn up of the cost of materials and £28 15s 7d was allocated. Eighteenth century bridges were no different from the Channel Tunnel today, and the actual cost of completion four months later had risen by a third to £38 15s 3d despite a last minute effort to save money by using the oak supports from the old bridge.

Fourteen thousand red bricks were brought from Halesworth and a special order placed for 48 coping stones. Skilled bricklayers were taken on and paid £1 8s per rod (about 4,500 bricks) together with two shillings in the pound beer money. In October the bridge was complete but George Simpson at Ubbeston Hall came up with an idea that he hoped would result in everyone getting their money back. He wrote to the then equivalent of the Rivers Authority suggesting that as the bridge was used by all and sundry, maybe all and sundry should pay for it.

The polite but firm reply came back, 'My very kindest regards to Mrs Simpson, but the whole burden of the Bridge must lay upon yourselves.'

In 1783 Joseph Hodskinson published a new map of Suffolk on which he showed some of the county bridges, such as Ballingdon, Sipsey, Langham, Southwold, Snape, St Olaves and the three Ipswich bridges, Stoke, Hanford (Handford) and Friers (Friars). But, as with John Kirby, the majority of bridges are unnamed and shown simply as the continuation of roadways over rivers and streams. Whereas the old fording places and subsequent bridges were of vital importance in ancient and medieval history, by the 17th and now the 18th century they had become part of the landscape and were unremarkable. If Hodskinson had been mapping Suffolk in Anglo Saxon or early medieval times, bridges and fords would have had top priority.

Some bridges, like the Black Bridge at **Bardwell** are famous just for being on Hodskinson's map while other larger and more prestigious crossings, such as those at **Beccles** and **Bungay**, are shown but not named. On the **Butley** to **Chillesford** road the bridge is named simply as 'stone bridge'. Littleport Bridge is clearly shown as is Prickwillow Bridge (both on the Cambridgeshire border), the latter crossing the River Lark (sometimes referred to as the Mildenhall Stream) a few miles west of **Mildenhall**.

Hodskinson, like Kirby, clearly marked the single-arched Bourne Bridge which forms part of the Wherstead parish boundary with **Ipswich** and carries the road from Ipswich to Manningtree (Essex). Quite how old this crossing is has never been fully explored but it goes back to at least the 1350s. The bridge, which has a stone arch with a projecting keystone, goes over Belstead Brook and the original small bridge can still clearly be seen as distinct from the late 19th century extension. This is one of the very few bridges left in Suffolk which has its original pedestrian refuges. There are also some peculiar rectangles set into the later parapets, presumably in imitation of the pointed originals.

The bridge was widened in 1891 and financed by the Ipswich Corporation, the County Council and private subscription. Alderman Nathaniel Catchpole of the Joint Committee performed the opening ceremony on 29th October 1891 accompanied by 'rural notables from the neighbourhood'. A worn plaque to this effect is still to be seen on the old bridge which is now pedestrianised and runs parallel with the modern one. But it is at least still there and was not demolished. Seats have been provided and it is a favourite place to sit and view the activities on the river, the yacht clubs and marina.

The bridge takes its name from the Manor of Bourne Hall but in the 1750s it seems to have been known as Bone Bridge. It appears in an advertisement that appeared in the *Ipswich Journal* in June of that year where the Ostrich Inn is said to be located at 'Bone-Bridge'.

It also appears as 'Bone Bridge' in the anonymous *Journal of a Tour through Suffolk, Norfolk, Lincolnshire and Yorkshire in the Summer of 1741.*

Bourne Bridge is the only site in the county where the rare, and poisonous, Corky-fruited Water-dropwort (*Oenanthe pimpinelloides*) can be found.

When topography and cartography first took off bridge names were sometimes included on maps for the first time but the choice appears arbitrary. Most bridges were unnamed although fords and early crossings would undoubtedly have once had possibly forgotten historical names. It is difficult to know if the names that do appear refer back to older appellations or were written down by 18th and 19th century topographers. In one stretch of the river between and Snape and Little Glemham there are at least four named bridges on the modern Ordnance Survey maps: Snape itself, Langham, Screw (railway bridge) and Beversham.

Rooks Bridge on the **Badingham** to Dennington road is also still on the modern maps and may refer to nearby rookeries. In the bridge vicinity the River Alde has a hard stony bottom reminiscent of a ford and is only a few yards off the route of a Roman Road. Does its name, therefore, derive from the ford or was it created the first time it was mapped? It is marked on Hodskinson's map.

Hodskinson also marks Apecol Bridge (or Apecot, it is difficult to tell) a short way north of Rooks but this is now barely discernible on the ground, the only hint being a Bridge Cottage by the side of the road. This may well be an example of a route that fell into disuse after the Romans left Britain in the middle of the 5th century.

Another bridge not far away over the infant River Ore, between Dennington and Framlingham, is Durrant's Bridge which was previously known as Durrance Bridge. It could also be the bridge known at one time as Dragon's Bridge. Nowadays it is barely worth a name at all but for whatever reason it was used by 18th century cartographers. The modern Durrant's could be thought to derive from a local family but given that it had an older name this is by no means certain.

The Frostley Bridge at **Dennington** over the Alde was built in the late 1700s but, although there are two keystones on both arches that might possibly have given the precise date, they are too worn to be read. On John Kirby's map this bridge is called Froizly Bridge which is curious, given that a 'froize' is a Suffolk word for pancake (as in the Froize Inn at Chillesford). The other possibility is that the name derived from 'froisy' which in the local dialect meant slippery, or difficult to handle. It could indicate that this crossing was in a frost pocket, hence frost(l) ey, or that the water regularly came up over the bridge and made it slippery and the name serve as a warning to travellers. Purists will point to 'froisy' being more commonly used in connection with loose corn but word usage over the years follows no particular rule.

On Hodskinson's maps it is 'Frostley' but elsewhere it appears as 'Frosley' and 'Frosty'. In the early 19th century probate and will of Francis Button of **Brundish** he gives to his eldest son 'two pieces of pasture with house, barn, orchard and yards, with the way leading from Frosty Bridge to Brundish Church.'

It is impossible, of course, to know the state of the bridges from Hodskinson's map but it is safe to say that not all of them had been consistently maintained over the years. Even town bridges, such as that at **Beccles,** suffered constant wear and tear, especially during the winter, and often received only temporary repairs. When the traveller Celia Fiennes saw it in 1697 the bridge was swamped in heavy rain just as it had been since it was built in the 16th century. She recorded:

> At the town's end one passes over the River Waveney on a wooden bridge, railed with timber and so you enter Norfolk, it is a low flat ground all here about so that the least rains they are overflowed by the River and lie under water, as they did when I was there, so that the road lay under water which is very unsafe for strangers to pass, by reason of the holes and quick sands and loose bottom.

Crossing the bridge sounds both dangerous and difficult. Celia Fiennes invariably relates the condition of the many bridges that she crossed and on entering Suffolk from Dedham in Essex she wrote:

> At this place I passed over a wooden bridge pretty large with timber rails of which make they build their bridges in these parts.

The bridge at **Bures St Mary** over the River Stour was in such a bad state that it was carried down river by high flood waters. It was reported in the *Ipswich Journal* on 6th November 1762:

> Bures bridge near Sudbury has been washed away by the violence of the flood on Tuesday last. Notice is hereby given that no wheeled carriages can pass over the Stour at that place. A new bridge will be erected as soon as possible.

The present Bures bridge links Bures St Mary with Bures Hamlet (Essex) and is one of the county's few cast iron constructions still in daily use. It was built in the 1870s by Rowson, Drew & Company of Upper Thames Street, London, and has a 20m span with 5 curved cast iron beams and brick jacks between. It is painted green and spans the river just below Bures church. There must have been a crossing here since at least the 9th century and probably long before, as Bures is one of the Suffolk villages said to have connections to the 10th century Saint and King, Edmund, who is thought to have been crowned here on Christmas Day, 855. Certainly there was a wooden bridge here in 1508, owned by the Priory at Stoke-by-Clare, which was then in a 'ruinous state'. In 1679 a masonry arch bridge was built and became known as Bures Great Bridge.

Another of several places where the old bridge was replaced by an iron one is at **Fornham All Saints.** Here the old wooden Causeway Bridge was badly damaged by the weight of a traction engine in 1890 and a man was killed when the timber gave way. The replacement by West Suffolk County Council became known as Sheepwash Bridge and appeared on the Ordnance Survey maps as such in 1955.

The perennial conundrum of who was responsible for bridge maintenance became, at the end of the 17th and into the 18th century, ever more contentious. For some the answer was clear: those who used the roads and bridges should pay for them. Turnpike houses and gates were set up on certain roads and travellers were required to pay tolls to use them. The money in turn was used to keep the highways and bridges in good order. Tolls had, of course, existed in various forms over many hundreds of years as had some turnpike bars but it usually depended on local initiative.

During the second half of the 17th century and for most of the 18th, Turnpike Trusts were gradually introduced. It is not within the scope of this book to explore the history of turnpikes but suffice it to say that bridges became a common and integral part of the emerging road network. Fourteen Turnpike Trusts were eventually set up in Suffolk. The first turnpike was granted in 1711 and ran from Scole Bridge (on the Norfolk side) to Ipswich.

Bridges were an incidental part of turnpike routes and played no particular role other than their being convenient places to cite a toll house as well as being, of course, a necessary component in the overall road system. At **Blythburgh** for instant, a former toll house is still to be found just north of the White Hart Inn on the approach to the bridge. The Blythburgh turnpike was not enacted until 1796 and came under the auspices of the Ipswich to South Town (Great Yarmouth) Trust. It is thought that spoil from the old priory buildings had been used over the years for bridge repairs and formed part of the new toll gate footings.

There was a turnpike at **Freckenham** Bridge over the River Kennett which is shown on Hodskinson's map of 1783. The Kennett marks the county boundary with Cambridgeshire and becomes the Lee Brook as it meanders through the village. The old bridge was demolished in 1954 and replaced.

One of the great legacies from the reign of Charles I is the restoration of the medieval Toppesfield Bridge in **Hadleigh** that was carried out in 1640. Jervoise described it as 'certainly the finest bridge in Suffolk, having three pointed arches each with six chamfered ribs. These, and the inner arch-rings, are built of stone, but the outer rings and the faces and parapets of the bridge are of brick.'

Constant maintenance is required if historic bridges are to endure. Work carried out on the bridge in the 17th century, and subsequently, ensured the survival of Toppesfield Bridge which was finally listed in 1950. This is the oldest bridge in Suffolk still carrying traffic.

The river crossings of the Stour, which carry traffic over the border between Suffolk and Essex, took on heightened political importance during the final year of the reign of Charles I. Three bridges on the Suffolk-Essex border, at **Nayland, Stratford St Mary** and **Cattawade** (sometimes called Cattaway)**,** saw action in the second Civil War.

The Siege of Colchester (in Essex) was a prolonged confrontation between the town's Royalists and the Parliamentarians, led by their commander, Lord Thomas Fairfax. At the start of the siege, in June 1648, the three bridges were considered crucial in cutting off any means of escape should any Royalists manage to break free from Colchester. Reinforcements, including Colonel Thomas Rainsborough, an expert in siege warfare, were sent to guard the crossings. Men congregated on the Suffolk side, anxious that the insurrection stayed in Essex. Fairfax was glad of extra forces and gave them leave to stand ready to defend the Stour should any of the Royalist leaders try to escape into Suffolk.

It was also vital that should anyone get free from the town to forage for food and supplies, they should be prevented from reaching the waterways. On 21 June a Royalist raiding party managed to get almost to the Cattawade Bridge where they summoned the garrison there to surrender. They failed in the attempt and the Suffolk contingent held the bridge.

Shortly afterwards some of the Royalist command attempted to ride out of the town heading for Nayland Bridge to take it by storm. But they were misled by their guide, possibly deliberately, and the alarm was given. Some fighting occurred at Boxted Mill (Essex) when the besieged royalists were again driven back from the Stour and prevented from entering Suffolk.

The siege dragged on and conditions inside Colchester's town walls became ever more dire, as food and ammunition dwindled to nothing. Sorties were less and less successful and few reached the three bridges still vigorously defended on the Suffolk side. On Monday, 28th August, Colchester surrendered to Fairfax who entered the town at two in the afternoon and ordered the Royalist colonels be shot the same day. The Suffolk troops that had secured the three bridges had done so with distinction and they were never breached.

It is likely that Boxted Bridge was then little more than a footbridge between Suffolk and Essex as there was no ancient road bridge at that point until one was built in 1788 by the River Stour Navigation Company. After 1895 responsibility for the bridge was shared jointly by the West Suffolk and Essex County Councils; in 1900 a steel girder bridge was built by Lexden and Winstree Union and is still in use.

The strategic importance of the Stour bridges was seen again, over two hundred years later, when the new bridge at **Nayland** was built in the 1950s. When dismantling the old bridge, workmen found a cache of dynamite underneath its timbers. It had been left there by the army during the Second World War. Had there been a German invasion the bridge would have been blown up for much the same reasons that the Suffolk men defended the bridge in the Civil War. Obviously the army had forgotten to retrieve the dynamite and for some years those using the bridge had been unaware of its presence.

There were at least three other bridges in Nayland. One was the small Pop's Bridge (thought to take its name from a family of that name) that crossed a stream south of the River Stour marking the flood relief channel that flows under the A134. There are records of the parish being ordered to repair Pop's Bridge in 1589, 1598 and 1649; it was finally washed away in floods during the early 1900s. The other was the Francis Bridge that crossed the Black Brook.

The timber bridge at **Stratford St Mary** had been in bad repair in the early 17th century and in the 1660s extensive renovation was carried out. Suffolk and Essex were jointly liable for repairs which were on-going as the bridge was constantly in danger of being swept away by floods. In the 18th century the Stratford bridge came to the attention of the writer and traveller, Daniel Defoe.

In medieval times drovers had had to use fords, as at **Combs Ford**, but by the 1720s they were using bridges and those fords which still existed alongside. Defoe, in *Tour Through the Eastern Counties,* writes about turkeys being driven over the bridge at Stratford St Mary. He came across a person who told him they had counted '300 droves of turkeys [for they drive them all in droves on foot] pass in one season over Stratford Bridge on the River Stour, on the road from Ipswich to London.'

These drovers, as they say, generally contain from three hundred to a thousand each drove, so that one may suppose them to contain 500 one with another, which is 150,000 in all; and yet this is one of the least passages, the numbers which travel by New Market-Heath, and the open country and the forest, and also by Sudbury and Clare, being many more.

The drives began after harvest, towards the end of August, and as well as turkey, geese travelled on foot to London, grazing as they went. Cattle, too, were driven down to London, one route passing through **Ixworth** and over the Hempyard Bridge. There was a cattle pound beside the bridge at **Walsham le Willows** though this could just have been for stray animals. The old bridge pound has gone but there is an 18th century one at the far end of the village.

It is not clear which of the **Cattawade** bridges Defoe was referring to when he wrote that when travelling from Harwich to Suffolk on the Tour 'I sent my horses round by Maningtree, where there is a timber bridge over the Stour, called Cataway Bridge.'

There was more than one crossing in the 1700s as indeed there is today. An old wooden bridge on the southern branch of the river was demolished in 1974 to make way for the modern White Bridge which carries the A137 over the River Stour.

At least three well-known Suffolk bridges were built or re-built during the reign of Queen Anne (1702-1714), that at **Mildenhall** (1709), the Temple Bridge at **Icklingham** (1711) and the graceful brick bridge at **Brent Eleigh** (*circa* 1713).

The original Brent Eleigh parapets have been extended over the years and later red brick piers are decorated with ball finials at each end of the parapets on either side. While it is in a beautifully shaded part of the village, spanning the River Brett, it is often very slippery due to the heavy canopy of trees that surround it.

Many of Suffolk's surviving bridges were built or rebuilt during the Georgian period. That constructed at the end of the reign of George II at **Chelsworth** over the River Brett is among the most picturesque in West Suffolk. In fact the bridge is not a single span but is a pair of narrow humped back bridges of red brick dating from the 1750s. Here is a mill stream and where there was once a causeway the local squire, Sir Robert Pocklington, caused the bridge to be built. There is a plaque on the upstream wall of the south bridge carrying Sir Robert's initials and the date that it was built. It reads simply 'R.P. Esq. 1745'.

It is classified as a Weak Bridge, is only 6ft 6ins wide, with a 3 ton weight restriction. Four sturdy wooden bollards guard its brickwork. Jervoise thought it to be 'the finest bridge in Suffolk, having three pointed arches, each with six chamfered ribs.'

Sir Robert Pocklington was knighted in 1794 with the Order of Maria Theresa by Francis II, the last Holy Roman Emperor, Sir Robert having rescued him in battle after Francis had been captured by French cavalry.

This bridge, like so many others, suffers cruelly from modern traffic. Repair work has been carried out on the piers of the south bridge and there is evidence of scraping and chipping of the brickwork by vehicles in spite of the bollards.

Chelsworth is a classic case where a watercourse was altered to accommodate a mill, thus altering the nature of the crossing places. The way the land was managed affected the volume of water in the river and streams. Changes in the landscape due to development and extended building, drainage and deforestation, increased run-off from the land and often put small bridges under great pressure. All this, of course, took place gradually and where mills were built it was necessary for the watercourse to be dammed which often destroyed an earlier ford or crossing point. Even by the Domesday Survey of 1086 there were nearly 200 mills in Suffolk and, therefore, as many mill streams and consequent bridges.

Since the 16th century the Mutford Bridge at **Oulton Broad** had been regularly washed away by storms and was in almost constant need of attention. In 1660 repairs were carried out that lasted until 1760 when the first brick bridge was built. This rigid structure meant that the small trading and fishing craft had to lower their masts to pass under it. But this bridge, too, was washed away and yet another was built: this one lasted until 1827. Successive bridges came and went until finally, in 1992, the lock was rebuilt and a pedestrian bridge incorporated into the new road layout.

During the 1770s and 1780s many of the county's bridges were rebuilt or heavily restored. The bridge between **Sapiston** and **Honington**, for instance, has the date 1780 carved on a stone in one of the bricks.

The construction of the bridge was completed only two years after the poet Robert Bloomfield began work at Sapiston on the farm of his uncle, William Austin, which formed part of the Duke of Grafton's estate. His mother kept Honington village school and it was from there that he set off to work in Sapiston in 1777, crossing the Blackbourn (various referred to as Black Bourne) stream on what was probably still an old wooden bridge.

The narrow hump-backed bridge now has a footbridge running alongside it and although the brick parapets are run-of-the mill, the attractive three arches and two cutwaters (one very recently repaired) can still be seen from the river bank where Bloomfield would have played as a boy. The bridge crosses the Black Bourne ('bourn' means stream) and there is still a discernible ford not far away.

The stream runs through Ixworth and **Fakenham Magna** to the Little Ouse at Euston.

Sapiston and Honington were used during the filming of *Things that go Bump in the Night,* an episode in the television series, *Dad's Army.*

The little wooden bridge at **Rushmere** (sometimes called the Henstead bridge) was replaced several times during the Georgian period but in 1846, Alfred Suckling recorded that 'the wooden bridge has already given way to a modern arch of brick which now spans the Willingham Water.'

Suckling also observed that in 1770 the bridge at **Herringfleet** was found to be in such decay that 'a new one was forced to be built in its place.'

The three-arched brick bridge at Fakenham Magna is dated 1848 and it is another of those where a pill box was sited next to the bridge in the Second World War. This little bridge is like so many others across the county, not grand in itself but a part of the landscape of Suffolk that would be the poorer without them.

At **Euston** there is a low parapet stone bridge over the Little Ouse which marks the boundary between Suffolk and Norfolk. Another Euston bridge is the graceful estate one, with stone panelled parapets, over the Black Bourn at Euston Hall, home of the Dukes of Grafton. It replaced a substantial late-18th century wooden bridge.

In 1778 the single-arch bridge on the **Walpole** to **Halesworth** road was built by subscription. Jervoise wrote that the date is carved on the keystone of the upstream parapet and it can still be seen.

The bridge in Walpole village is 20th century having been reconstructed in 1939. Its 'arch' is a straight section of ribbed concrete.

Other bridges were built during the long reign of George III, including that at **Baylham** which is dated 1780. There is a raised footpath for when the River Gipping floods and the two little hump backed bridges, so very reminiscent of a bygone age, are limited to wheeled vehicles under 8 tonnes.

The red brick **Glevering** bridge on the Easton road was built in 1777 and is Grade II listed. It crosses the River Deben and the date is noted on the keystone of the main arch (there are smaller arches either side). Half of this bridges lies in the parish of Hacheston. John Kirby passed over an older Glevering Bridge in the course of his travels in 1735 but, as usual, it was just a way marker and he gives no information or description of it.

However, since he lived at the water mill in **Wickham Market** Kirby would have known the bridge over the River Deben as it was in the 1730s. This crossing has seen many generations of bridges and those of the 20th century featured on many postcards of the time. The pretty little bridge with criss-crossed parapets, with turned wooden uprights, came to grief in 1912. On 8 November the bridge collapsed into the river reputedly as the result of damage by a steam engine that belonged to the Rackham family who owned and ran the water mill.

At the end of the 18th century there had been no significant bridge legislation since Henry VIII's Statue of Bridges in 1530. The **Bridges Act** of 1803 gave more authority to those officials engaged by the Crown to be responsible for the upkeep of bridges and the roads, principally the County Surveyor. They had greater jurisdiction not only over the bridges but for roadways 100 yards past the ends of the bridge.

Little Bealings white brick bridge over the River Fynn

Bridge at Bures St Mary on the Suffolk-Essex border

The Rendham-Sweffling bridge over the River Alde

Stowmarket's Navigation Bridge beside the Old Maltings

Homersfield Bridge on the Suffolk-Norfolk border

The Red Hand of Ulster on Homersfield Bridge

Sir William Cubitt's cast iron bridge at Brent Eleigh

The Priory Bridge at Clare

Plaque on the Steggles Bridge at Bury St Edmunds

The White Bridge on the Suffolk-Essex border at Cattawade

*River Gipping at Sproughton Mill, the wooden-railed road
bridge in the background*

The lattice bridge at Santon Downham

Ordnance Survey Bench Mark bolt on Snape Bridge

Note Ordnance Survey cut mark on Beccles Bridge over the River Waveney

Great Bealings bridge over the River Lark undergoing repairs to its parapets

Rushford Bridge on the Suffolk-Norfolk border

CHAPTER 2

During the 18th century the Industrial Revolution gathered pace. Thus far bridges had been a means whereby to pass safely and dryly over a natural watercourse. From 1705 onwards Suffolk was opened up by various navigation systems and although it was still a combination of water and bridges, sometimes incorporating existing bridges but often new builds, rivers were diverted and waterways straightened. All of a sudden Suffolk had a new riparian landscape that consisted of staunches, locks, wharves and new bridges of varying size and construction.

The **Stour Navigation** was opened in 1705; the **Blyth Navigation** opened in 1761; the **Gipping Navigation** to Stowmarket was completed in 1793; and the **Lark Navigation** began around 1715.

The Stour Navigation went from Sudbury to Manningtree in Essex and was one of the country's first such schemes. However, because it was such an early navigation there was no example to follow and no advice to be had; therefore, although an Act of Parliament made the river navigable, no one thought to secure agreement for horses to travel along the tow paths. Consequently the existing tow paths often switched from one side of the river to the other which meant that the horses had to jump onto floating platforms or strengthened foredecks and be poled over the river to continue the journey. Sometimes the horses were required to jump over fences that were erected down to the waters edge to prevent livestock from straying to a neighbouring meadow. There were around 123 fences between Mistley and Sudbury and records show that many had to be cut down to around three feet to enable the horses to pass over them.

One of the famous 'Stour' paintings by John Constable is entitled *The Leaping Horse* (1825) and shows a rider urging a barge horse to jump over a barrier on the towpath.

Many new bridges had to be built along the navigation, mostly to replace fords, and the early bridges were simple, timber-built structures which generally followed a common pattern.

The **Blyth Navigation** ran for 7 miles (11km) from **Southwold** to **Halesworth**. It was opened in 1761 and lasted for over 120 years, becoming insolvent in 1884 when matters were not helped by the attempt to reclaim saltings at Blythburgh which resulted in the estuary silting up. It was used occasionally until 1911 but finally abandoned altogether in 1923.

The bridge at **Mells** is one of the most interesting of those along the Blyth Navigation and has modern relevance in that there is now a sluice and gauging station nearby. The bridge was so badly damaged by heavy traffic that in 2016 work was carried out to remedy the bulging sides which were protruding some two feet either side of the bridgeway. The adjacent Holton Flood Arch and nearby Railway Bridge was similarly strengthened at the same time.

The hump backed navigation bridge at **Blyford** was entirely unsuited to the build-up of traffic during the 1940s and 50s. Also known as Blyford Lane Bridge it was demolished in 1963 and a concrete bridge substituted; what seem to be the footings for the old rails leading off the brick parapets are still visible in the hedgerow. Similarly, the somewhat later **Wenhaston** bridge, only a hundred yards from the Blyford one, spanned the old course of the River Blyth. This ran alongside the old ford, known locally as the Wenhaston Run, which was the original river crossing point. The bridge was a pretty railed affair with two piers at the ends of both sides. The ford would have been the site of the traditional 'causey', or causeway, which featured in Richard Pepyn's will of 1469 in which he leaves money to repair the causeway 'betwyx Wenhaston and Blyford'. A new road system was built in the 1960s obliterating both bridges and the ford.

The **Gipping Navigation** ran from just north of **Stowmarket** to Handford Bridge in **Ipswich** and was opened in 1793. Tolls were set at 1d. per ton per mile downstream but only 1/2d per ton upstream and there was a minimum charge equal to a 35 ton load. Manure from the London streets was carried upstream and, for whatever reason, travelled toll free.

Variously called the Gipping Navigation or the Ipswich and Stowmarket Navigation, it had a total length of 16 miles with over 15 locks, several of them can still be seen today. The Gipping Way is a public footpath that runs the length of the navigation and affords the opportunity to see the bridges and weirs along the old route.

At **Stowmarket** a variety of bridges awaits, not least the three main ones in the town: the Pickerel Inn bridge, the wide single arched brick bridge near Navigation Approach, and the modern bridge near the industrial estate that carries a series of murals depicting the town's history.

There are also unusual vehicle parapets on the Stowmarket Relief Road bridge.

The Navigation Approach bridge is beside what is left of the Stowmarket Maltings complex, which opened in 1793 as one of the beneficiaries of the navigation. In an 1838 drawing of Stowmarket Quay by Henry Davy, the river is traversed by an unsteady-looking footbridge which marked the limit of navigation for the barges and river traffic.

At **Needham Market** the route of the navigation is discernible under a sturdy concrete bridge built by George Mundy & Sons of London. It was financed by East Suffolk County Council in 1922, towards the end of the navigation era.

The next bridge downriver is that at **Creeting** (previously known as Lower Bosmere Mill), which has been restored by the River Gipping Trust. Work began in 1994 here and at Bosmere Lock and took ten years to complete.

After Creeting Lock is **Pipps Ford** where the old navigation bridge can still be seen. There is also a new footbridge, known as the Mathematical Bridge, just the other side of the double gates on the by-wash (on private land). The oak for the footbridge was felled at Great Glemham and the design follows that of the Mathematical Bridge in Cambridge. It was said that the original 18th century Cambridge version was constructed without nuts, bolts or nails but this is a fallacy, as is the story that the bridge was designed by Isaac Newton. Although the bridge appears to be an arch, it is composed entirely of straight timbers but built to an unusual and complicated engineering design which gave it its nickname.

One of the best surviving examples of an 18th century navigation lock, once part of the Gipping Navigation, is that at **Baylham** where there are also two bridges and a raised walkway beside the old mill. One of the bridges is hump backed and built of red brick with 2 semi-circular arches and a third smaller arch alongside.

Immediately north of the present bridge is the site of a Roman settlement, Cambretovium, and the crossing is likely to have been forded or even bridged by the Romans since the settlement included two forts and a large civil presence. Part of the Cambretovium site is occupied by the Baylham House Rare Breeds Farm.

The lock at **Claydon** was lost in the construction of the A14 trunk road and the current bridge crosses the river at **Great Blakenham.** An etching in the British Museum executed in 1837 shows a wooden bridge supported on four spindly supports to have existed in Claydon, whether at the site of the modern bridge or not is unclear. John Kirby mentioned a bridge at Claydon in 1735 and it is likely to be this one. It portrays the rural idyll to perfection, the river idly flowing under the bridge, sheep grazing on one bank and a figure standing beside a line of wooden stock railings on the opposite bank and a large oak tree beside a footpath leading to the bridge.

Perhaps it was this bridge referred to by John Longe Vicar of Coddenham, who wrote in his Diary for 19th December 1831:

I attended a trustee meeting [of the] Claydon Turnpike at Claydon. Tolls not let. Mr Brown and I ordered repairs to the railing, etc., of Claydon bridge to be done at the expense of the county.

Claydon was on the turnpike that ran from Ipswich to Scole and also on that to Stowmarket and Haughley.

The lock at **Bramford** was the fourth up-river from the Handford Sea Lock and until the floods of 1939 also had an iron bridge which had replaced the old wooden one in 1904. The tale is told that in 1939 a group of school children were to be taken across the bridge in an old tumbrel as the water began rising to dangerous levels and people were anxiously waiting to cross the river. The horse, named Prince, managed one return journey but by then the flood water had reached his belly and the hump backed bridge over the stream was lost to view. Prince stopped and refused to make a third crossing and would not set foot on the bridge. A minute later the brickwork supporting the bridge collapsed and within a few more minutes the whole structure had disappeared under the raging water. Prince was hailed as a local hero.

Within two days a small wooden footbridge had been slung across the river and some months later the Army erected a Bailey bridge that stayed in place for several years.

At **Sproughton** the scene is idyllic but along the footpath, passing the graceful footbridge to the mill, there are echoes of that hustle and bustle in the noise and speed at which the water still gushes over what was the old lock. The Gipping itself is still discernible, trickling along beside the navigation channel. The original river is said to have been navigable in the first century when the Danes used it in 860AD to establish a settlement at 'Ratles-Dana', now Rattlesden.

Sproughton's current road bridge has angled concrete girders but wooden railings at road level that are sympathetic to the surroundings.

The **Lark Navigation** linked Bury St Edmunds commercially with Kings Lynn, Ely and Cambridge between 1715 and 1855 but this 'water highway' as it was called was left in a sorry state when the navigation owner Sir Thomas Gery Cullum died in 1855. Its demise coincided with the growth of the railways but there were also problems with the rights of navigation themselves having been sold to absentee proprietors. In 1889 the Marquis of Bristol and the mayor of Bury St Edmunds purchased the River Lark Navigation and reopened the waterway to traffic.

There are some twenty-four bridges, locks or sluices along the route of the Lark

Navigation which ended, or began, at St Saviour's Wharf, Bury St Edmunds. Those using the waterway often found it difficult negotiating the various staunches and it took time and money to bring the navigation into a useable state again.

Along its route stood Temple Bridge at **Icklingham** which is now known as Temple Staunch although remains can be seen of the old bridge and further up river is the Farthing Bridge where evidence of the navigation can also be found. A little way from the bridge, in the middle of the river, is what appears to be an ancient pier possibly belonging to an earlier crossing.

As late as 1899 barges were still traveling up river to Temple Bridge, where a stone quarry was opened, and on the Whitsun Bank Holiday Monday of that year the journalist and writer, William Howlett, counted over 40 boats between Temple Bridge and Mildenhall, most of them laden with day-trippers. Two years earlier he had ferried up to see the new bridge at **Judes Ferry** where he observed a long gang of barges going through the Wamil Stauch.

In September 1900 William Howlett made the journey from **Barton Mills** to the bridge at Prickwillow (Cambridgeshire) in his boat *Annie* with the aid of a towing donkey. When they arrived at Kings Staunch it was necessary for the donkey to cross the narrow bridge which, observed Mr Howlett, was in a very bad state of repair. The donkey was of the same opinion since no amount of encouragement by his handler Robert Peachey persuaded it across the bridge:

> Carrots were held out to him, and coaxing, and anything we could think of, but NO, 'moke' would not cross the bridge. We fixed ropes on him to try and pull him over, but he laid down perfectly flat, and after an hour's trial to get him over, with no success, we left him tied to a post until we returned - and we had to take it in turns to tow the boat ourselves!

The Prickwillow Bridge had only recently been built (in 1866) and this girder bridge is the last crossing over the Lark before it joins the Great Ouse.

At the start of the 19th century, in spite of all the commercial activity involving bridges by both the navigations and turnpikes, there was a small resurgence of the old paternalistic habit of private money being used for the good of the community and it was fashionable for benefactors to leave both their names and that of the surveyor or builder on the bridge itself.

In **Hoxne** the Kerrison family, in medieval style, gave land for a village hall and funds to rebuild Goldbrook Bridge.

Sir Edward Clarence Kerrison inherited the baronetcy of Hoxne and Broome from his father who had served in the Peninsular War and as Colonel commanded the 7th Light Dragoons at the Battle of Waterloo. The title was created in 1821 but died with the childless Sir Edward in 1886.

In 1878 Goldbrook Bridge was rebuilt and the following year the Parish Reading Room (now Village Hall) was built and included a roundel of St Edmund hiding under the bridge. When the new footings were being dug a number of ancient artefacts were found including an iron spearhead and horseshoes.

A few years earlier in 1858 the Swan Bridge was also rebuilt. There are sandstone plaques set into the walls naming the surveyor but they are too badly corroded to read.

Among the few bridges on which pedestrians and cyclists can stand and stare is that at **Homersfield,** now closed to vehicular traffic. It crosses the River Waveney and straddles the boundary between Suffolk and Norfolk, standing partly in the civic parishes of Homersfield (Suffolk) and Aldburgh and Wortwell (Norfolk). Seven administrative boundaries meet in the centre of the bridge.

Built in 1869 for Sir Robert Alexander Shafto Adair of Flixton, it is a rare monument to bridge engineering and had the distinction of being awarded a Grade II listing in 1981. The wrought iron framework of the arch is encased in concrete and as such is the oldest surviving concrete bridge in Britain. Embedding the iron structure in concrete was revolutionary at the time and laid the foundations for the engineering process that became known as reinforced concrete at the end of the 19th century.

The bridge was designed by the architect Henry Medgett Eyton, then living in Ipswich, who employed W & Thos Phillips of London to carry out the work. In December 1869 Henry Eyton reported back to Sir Robert (known locally as Sir Shafto) that he had approved Phillips' quotation of £344 which he thought 'a low price compared with some of the County bridges.'

The bridge has a single arch span of 48ft (15.2m) and the cast iron balustrades, currently painted dark green, are each decorated with the Adair monogram.

In 1970 traffic on the A143 was diverted to a new bridge thereby providing essential respite to the concrete fabric. Some years later the ironwork was found to be badly corroded and some of the lattice framework exposed in places where the concrete had fallen away. In the 1990s the bridge was compulsorily purchased by Norfolk County Council and restored with funding from eight separate entities, including Waveney District Council, English Heritage, Suffolk County Council and Blue Circle Cement. It was opened as a footbridge and cycle path in 1996.

Sir Shafto's bridge replaced that of 1763 built here by William Adair, Lord of the Manor of South Elmham, who bought the estate containing Homersfield in 1753. Adair had the right to levy a toll on anyone passing over the bridge and another for the return trip. Perhaps the Ghost of Homersfield Bridge was one of those who thought it should be possible to obtain a reduced rate for the return trip, as operated at some other bridges. The ghost apparently had lain under the bridge for a very long time but in the course of repair work was released and can be heard moaning and howling when his spirit was freed.

Some of the moaning and howling could also be attributed to a poor ostler who was reputedly beaten to death at Flixton Hall. Before he died he left bloody handprints on the walls which is said to account for the four bloody hands on the Adair crest that is mounted on both the up and downstream spans. Local legend has it that the Adair coat of arms itself only had three hands and a fourth was added to remind the family of the dreadful deed.

The true explanation is that the Adair family had lived in Ballymena, County Antrim, since the 1620s and took as their coat of arms three of the 'Red Hand of Ulster' with a bearded head above a shield. The head is couped and spilling blood, which is said to commemorate some ancient encounter with an enemy chieftain. The Irish generally cut off the heads of chiefs they had slain in battle as they considered no man to be dead until his head was cut off. Thus the shield became known as the Bloody Hands of Ulster.

Originally the shield had three dexter hands (that is, the wearer's right side and the spectator's left) but the one on Homersfield bridge has four red hands. Careful examination will reveal that the centre hand is shown as sinister (wearer's left and spectator's right). This is not to commemorate the murdered ostler but in heraldry a sinister red hand was used to indicate a baronetcy, in this case that of Sir Shafto Adair, also known as Lord Waveney.

Technical advances of a different kind can be seen at **Culford** where an iron bridge of 1803 spans Culford Lake (nominally the River Cul) in a park landscaped principally by the renowned landscape designer, Humphry Repton, in 1791 when he produced one of his famous Red Books. Repton's design followed a 1742 landscaping plan by Thomas Wright; the bridge was designed in 1798 to a patent lodged earlier by the architect Samuel Wyatt and commissioned by Charles, 1st Marquis Cornwallis, owner of the Culford Estate from 1762 to 1805.

Samuel Wyatt had a diverse and distinguished career and was a member of a leading family of 18th and 19th century English Architects. On 10 June 1800 he was granted a patent for the use of cast iron and 'for his invention of a new Art, or Method of making and constructing Bridges, Warehouses, and other Buildings, without the use of wood, as a necessary constituent part thereof.'

The 60ft span bridge is Grade I listed and is the fifth oldest bridge of its type in the world. It formed part of a new roadway leading to the Hall from Newmarket Lodge, a pleasant drive through the park for Charles and his guests on their way to or from Newmarket Races. The curved iron structure of the bridge was cast from cannon that Charles had brought back with him from India where he served two terms as Governor General.

Original plans for the bridge were lost between 1798 and 1803, when they resurfaced. Advice was sought from William Hawks of Gateshead and under his supervision the iron matrix of the bridge was cast.

The bridge at Culford is the earliest known example that has six tubular. i.e. hollow, ribs which were cast by William Hawks. Eighty tons of iron castings were used but the bridge has masonry balustrades and marble urns. Hawks erected the bridge, with its stone balustrades, and it survives in its original form.

Since 1935 the property has been home to an institution that had begun in Bury St Edmunds in 1873 called the East Anglian School for Boys but renamed Culford School. It is now an independent day and boarding school. Conservation work was carried out on the bridge in 1998, when sandblasting and repainting took place. It can be seen from the nearby public footpath.

The idea of building bridges from cast iron soon caught on and William Cubitt (later Sir) was employed by the burgeoning Ipswich firm of Ransome & Son in 1812 to take charge of iron casting. Born in 1785, Cubitt was the son of a Norfolk millwright and as well as being a civil engineer, invented self-regulating windmill sails; he was the chief engineer of the Crystal Palace that was erected in Hyde Park in 1851 and became president of the Institution of Civil Engineers in 1850.

Cubitt's first cast iron bridge, now a scheduled Ancient Monument, crosses the River Brett at **Brent Eleigh** beside the Hadleigh to Lavenham road. It has a single span of 13ft and was constructed in 1813. When the road was bypassed in 1953 the little bridge was left more or less intact. It sits unnoticed beside a busy replacement highway and is now an almost invisible relic of Suffolk's important role in the national prosperity of the Victorian industrial era.

Sir William's second cast iron bridge is at **Clare**. It carries a minor road over the River Stour and is named the Priory Bridge. It was also built in 1813 with three semi-elliptical spans of 11ft (3m), 13ft 5in (4m) and 11ft (3m). Although it was built at a time when Clare was on the main route from Bury St Edmunds to London, it survives the pressures of modern traffic and is still in use today although its egg shell blue paint is peeling. It is more elaborate than the bridge at Brent Eleigh and the west face has an extra cast iron beam to widen the south approach.

The cappings carry the mark of Ward & Silver of Melford who worked on the bridge in 1869. Although there is no plaque on the structure (except the 1813 date) it is thought the iron was cast at Ransomes of Ipswich who were to be in the forefront of design and supply of the railway system and many other projects that were to come in the 1840s.

Another very early use of cast iron is to be found at **Helmingham** where a pair of ornate bridges span the 60ft wide moat around Helmingham Hall, home of the Tollemache family and the largest moated house in Suffolk. Extensive alterations to the hall were carried out between 1800 and 1803 from plans drawn up by the Regency architect, John Nash, who envisaged very delicate and intricate bridges. Perhaps their design was too complicated as the bridge plans were dropped until some years later when the present ones were built.

There is no doubt which firm made the bridges as 'Ransome Ironfounder Ipswich' is found on the top flange of the bridge girders. The bridges, one to the south-east and the other to the north-east of the hall, retain working drawbridges, which were originally operated with a windlass but are now worked by an electric motor. The drawbridges have been raised every night at Helmingham since 1510.

In the north-west corner of the grounds is a small brick and flint bridge built around 1815 that crosses a stream at a point known as The Dell. In around 1801 John Constable painted several versions of *A Dell in Helmingham Park.* In a sketch that he made in 1800 the bridge looks to be a primitive and rickety wooden span, albeit picturesque. Constable's brother was steward of the Tollemache woodlands and John Constable lived for a time at Helmingham Rectory. It is believed that the oak tree still standing in the Dell was the one sketched by Constable.

In 1990 the aforementioned cast iron bridge over the River Stour at **Bures St Mary,** on the Suffolk-Essex border, was found to have cracks in some of the beams and the bridge reduced to one lane with a 7 tonne weight restriction. In due course repairs were carried and the bridge, built in the 1870s, afterwards reopened to full weight traffic with the overall appearance unchanged.

Another of Suffolk's 19th century iron bridges is on the Suffolk-Essex border not far away from Bures at Wormingford (Essex) where the emblems of the two counties appear on the respective ends of the bridge: the castle, or portcullis, for East Suffolk County Council and three Seaxes 'fesswise' for Essex. Wormingford's Church Road bridge replaced the old ford in 1802, still called a horse bridge in 1812, and in the 1820s a new wooden one was built. This collapsed during the winter of 1895-6 and replaced by an iron bridge in 1898.

Until the start of the 18th century most bridge building had been, by and large, anonymous. No one had laid claim to having built or designed a bridge; there were no plaques to priories or abbeys or even to private benefactors on the keystones. They were built anonymously according to the technology of the day, the particular terrain or watercourse it had to span, and what funds were available. There had been a few who left a plaque for posterity, such as William Abell at **Nayland** and Robert Pocklington at **Chelsworth** but over the next couple of hundred years names of companies, manufacturers, borough or council mayors, became more common. Manufacturers and foundries understandably wanted to put their trademark where everyone could see it.

The building firm of Steggles left their stamp on at least two Suffolk bridges, the first at **Cosford**, where in 1840 a plaque notes that the bridge surveyor was William Steggles and J Corder the builder.

The second is on a plaque on a **Bury St Edmunds** bridge, built of white brick, which reads 'Steggles & Son, Builders 1840'. The Steggles Bridge spans the River Lark and carries a huge amount of traffic into the town centre. Its predecessor was pulled down in 1838. Iron railings were erected along the river bank, reused from the Market Cross, and the Marquess of Bristol subscribed to the work. When the bridge was constructed it did away with the ford which, until that time, had always existed beside the bridge in Eastgate Street. One of the earliest bridges on the site was there in 1148 and the mediaeval bridge chapel dedicated to St Mary stood nearby.

The Victorian era brought with it an entirely new concept of a crossing, the railway bridge. This needed engineers who had reputations to make and keep. For the first time people travelled under bridges, on one or other navigation, almost as much as they had previously travelled over them. After 1846 when the first railway was opened in Suffolk this new type of bridge proliferated and soon altered the county landscape. There was no bridge evolution to these new crossings, as there had been when ford became wooden bridge and then stone, and they anyway crossed not only rivers, navigations, marsh or fenland but also roads. These were entirely new types of structure, built for a specific purpose. These were no packhorse bridges with a ford besides where trains of horses, farmers with carts on their way to market, or those on foot going about their daily business passed.

In 1845 an Act of Parliament was obtained to begin building a railway from Ipswich to Bury St Edmunds via Stowmarket. It was the coming of the railways that accelerated the development in **Ipswich** from the Cornhill down to the River Orwell during the second half of the 19th century.

One of the most significant 19th century river bridges to be built in **Ipswich** was the Princes Street Bridge that brings traffic down from the town to the railway station. It was part of the grand development of the town following the opening of the railway in the mid-1800s and the plaque on the present bridge says it all:

> This bridge was constructed by the Ipswich County Borough Council & completed in 1927 replacing one of wood and iron erected by the Eastern Counties & Eastern Union Railway Company in 1860 when this road was first opened.

Princes Street runs south from Cornhill to the bridge making it one of the longest streets in the town. The bridge had been proposed by the Railway Company in 1845 but the street was developed in two halves as there were prolonged negotiations with property owners along the route so it was not until some years later that the bridge became a reality.

At the same time as the Princes Street bridge was under construction, work was also being carried out on a new bridge at the Stoke crossing over the Orwell. The plaque on the concrete semi-circular parapets reads:

> Erected AD 1924-25 – Engineer Sidney Little, A.M - Boro' Engineer & Surveyor – Contractors D G Somerville & Co Ltd.

When it was built many timber framed buildings were demolished, including the White Lion public house that had closed in 1913.

The bridge it replaced had been an iron one built by Sir William Cubitt in the 1890s and a generation on from the stone bridge that was partially swept away by floodwaters in 1818 when three men standing on it were drowned. Cubitt, as chief engineer at J & R Ransome, had sourced the cast ironworks from Staffordshire and shipped them to Ipswich from Lincolnshire.

A second Stoke Bridge was built alongside the 1920s one in 1983 to cope with the ever-increasing burden of traffic. Both bridges have pedestrian walkways on the outer parapets.

The proliferation of railway bridges ushered in a new era where stations were built along with the inevitable Station Hotel, or Station Inn, to cater for travellers waiting for their trains or looking to break their journey. This mirrors the many river crossings which over the years spawned a number of inns either side of bridges and fords, especially if the waters were tidal and travellers had to wait for the water to recede.

Throughout the centuries, traffic at the Stoke crossing in **Ipswich** was well provided for in the way of inns and taverns, and in the 19th century men with tumbrels and wagons found it convenient to patronise the 'Old Bell Inn' that still stands opposite the bridge on the Stoke side. There has been an inn at this crossing since at least 1639 as it is mentioned in the town assembly book as already existing in that year.

At the **Wherstead** end of the Bourne Bridge on the south bank of Belstead Brook stands the 'Ostrich' which is thought to have been built in around 1612. In 1888, Dr J E Taylor wrote:

> Fifty years ago the 'Ostrich' was a favourite Ipswich trysting-place in summer. The gardens were then famous. They are still well kept and pretty, and are much visited on summer events. The 'Ostrich' and Bourn Bridge form a favourite bit for artists.

Dr Taylor records that there were 'inscriptions of people on the small panes of the window facing the river – some of them being a century old.'

The inn took its name from the ostrich in the crest of Sir Edward Coke who acquired the Manor of Bourne Hall in 1609 and a year later he bought the land 'in the river channel or crecke in the west part of Borne Bridge' and built or rebuilt the inn. Its name was later changed to the 'Oyster Reach'.

A tale is told about an incident that took place at Bourne Bridge in 1803 when a Mr Scott, head gamekeeper to Sir Robert Harland of Wherstead Lodge, rode over the bridge to the Ostrich Inn where he tied his horse to the hitching rail and proceeded to order a drink. However, there were two known members of a poaching gang in the bar and from their threatening behaviour Scott guessed there was trouble brewing. He followed the men down to a nearby field where gang members were armed with guns and bludgeons. Mr Scott hastened to inform Sir Robert of the situation and it so happened that General Lord Paget, commander of the Light Brigade of Cavalry, was dining at the Lodge. The Light Brigade was then stationed at Ipswich and Lord Paget sent order to the cavalry barracks that a detachment was to be sent to Bourn Bridge without delay.

Meanwhile back at the Ostrich the poaching gang saw the game was up and tried to escape over the bridge, but their way was barred by the dragoons who easily held the narrow bridgeway. Men were positioned in the other exit roads and the corporal in charge stationed himself immediately outside the Ostrich in the centre of the Manningtree road. Two of the men tried to escape via the river but were easily caught, as were the other gang members, and the dragoons restored order to the proceedings.

In 1891 Bourne Bridge was expanded to give a roadway of about 30 feet in width. The sandstone plaque is now very worn but declares the widened bridge to have been built by the County and Borough Councils and private subscription. It was opened by Alderman Nathaniel Catchpole, Chairman of the Joint Committee, on the 29th October.

CHAPTER 3

It is not unusual to find First or Second World War pill boxes beside Suffolk bridges. The threat of invasion has loomed over the Eastern Counties since Roman times. The numerous waterways left the county vulnerable during the 11th century Viking raids and it is said, but so far not proven, that bridges were erected over small rivers sooner than might otherwise have been to try and prevent ships from penetrating inland. Mights Bridge at **Southwold** has a long history as a potential barrier to maritime incursions and in the years leading up to World War One it was seen as a possible invasion point.

Mights Bridge 'Tumble Down' pillbox was one of the first to be built in order to guard the bridge, Buss Creek and all roads inland. It is circular and a very rare example of its kind. Southwold, of course, was no stranger to warfare and had witnessed sea battles, not least the famous Battle of Sole Bay in 1672. Then there were the proposed fortifications in Tudor and Elizabethan times and in 1796, during the Napoleonic Wars, town life was against disrupted when two volunteer corps were set up. As it happened no invasion took place either then or in 1914 when the Germans believed that the old canon guns visible on the cliff top were active.

During 1940 many more pill boxes were built elsewhere at the ends of several bridges. There are two in close proximity to the bridge at **Rushmere**, another near the three-span humped back bridge at **Fakenham Magna** and a small one close by **Euston** bridge over the River Little Ouse on the A1088. There is one near to Reckford Bridge at **Middleton**, one near **Wilford** Bridge on the River Deben, one at **Mendham** and another at **Homersfield**. The pill box at **Bures St Mary** stands at the bridge end on the Essex side of the river.

The bridge at **Judes Ferry** over the River Lark on the south side of **West Row** was considered to be of strategic importance in 1940. Pill boxes were constructed all along the river and the bridge itself was mined, the bridge area being a designated 'check point'. The bridge was defended by two roadblocks and to the south east stood an anti-tank gun emplacement. By 1942 the immediacy of the situation had lessened somewhat and the roadblocks were manned by the 2nd Battalion Cambridgeshire Home Guard.

Jude's Ferry Bridge was part of the 18th century River Lark Navigation and today is the limit of navigation. In 1994 Suffolk County Council inspected the old 1890s bridge and in 1999 a new bridge was built.

Some years later another Suffolk bridge became the focus for the Home Guard but this time the platoon members were actors. The single-width bridge at **Santon Downham** featured in the television series *Dad's Army* which was screened in December 1972. Onlookers saw the programme's characters Captain Mainwaring, Sergeant Wilson, Corporal Jones and Private Godfrey in a fire engine crossing the diamond-patterned iron bridge on their way to a fire. The *Brains not Brawn* plotline called for the platoon to disguise themselves as firemen in order to fool any passing Germans but were, in the event, diverted to a 'real' fire.

This white-painted iron structure is imaginative and striking in design but suffers like other metal bridges of its time, in that its surface is dulled in places by algae. The brick piers on either bank are of Thetford white bricks in English bond and the bridge itself strengthened by iron tie-beams. It was built originally as a rail bridge at the start of World War One by the Canadian Army, to get timber and logging machinery over the river. The rails were lifted in 1919 when the estate of which it was part was sold to a land company; in 1923 it was sold on to the Forestry Commission and eventually it became a road bridge. In 1992 it was proposed to replace the bridge with a more modern structure but the plan was opposed by the locals and so the lattice bridge survives.

In **Thetford** (now in Norfolk) there is a sculpture of 'Captain Mainwaring' sitting on a bench close to the Town Bridge together with a Dad's Army Museum in the town.

At **Ixworth** reminders of World War Two could be seen around the village for many years after the cessation of hostilities. Concrete cylinders were left behind by the army and the local council had no idea what to do with them. They were eventually employed at either end of the three-arched Hempyard Bridge to help prevent erosion: while they have slipped quite a bit over the years they can still be seen at the bridge ends. Hempyard Bridge is described as a packhorse bridge and crosses the Black Bourn although its origins are probably to be found in the nearby Augustinian Abbey. It is said to take its name from the 17th century hemp yards that lay to the west of the river but there was a considerable hemp industry here and at nearby Walsham le Willows in the 14th century.

Ixworth was designated a Defended Place by the Eastern Command Defence Plan and one of the many pillboxes was built beside the entrance bridges. The village was encircled by barbed wire with weapons pits and machine gun emplacements constructed to enable Spigot Mortars to be fired as anti-tank weapons.

Ballingdon Bridge at **Sudbury** also had an anti-tank gun positioned close by. In a photograph of the concrete bridge erected in 1910, one of the balustrades is missing thought to have been removed in 1939 or 1940 to give the gunner a clear field of fire. Had the German invasion reached the bridge it is likely it would have been blown up.

In the early years of the war a very unpopular decision was made by the Home Guard at **Southwold** when all but one span of the narrow-gauge railway bridge that crossed the River Blyth to **Walberswick** was blown up. The 146ft long swing bridge turned on a steel caisson sunk in the river bed and was the pride and joy of all who travelled or worked on the famous Old Southwold Railway. After several false starts the line from Halesworth to Southwold was given the go-ahead by Act of Parliament in 1876. Sleepers were ordered from Norway and rails from South Wales and the track opened in 1879. It ran alongside the Great Eastern Line at Halesworth before branching off to Wenhaston and then Walberswick and so to Southwold across the river bridge.

Local legend has it that one of the trains was acquired from Peking having originally been built for the Emperor of China. The writer W G Sebald became fascinated with the bridge and its Chinese connections and wrote extensively about it in his novel, *The Rings of Saturn*. He makes the point that there was a British armed presence in China during the Opium Wars of 1840 but was unable to discover how 'this diminutive imperial train' intended to connect the Palace in Peking to one of the summer residences, ended up on a branch line of the Great Eastern Railway.

It would seem that the train, if it did indeed come from China, was probably ordered for the Emperor Kuang-hsu but the order was cancelled in the mid 1890s when the young Emperor began to espouse the reform movement. How it got to Suffolk has not been discovered but the story persists.

After the war ended missing sections of the demolished bridge were replaced by a temporary Bailey bridge and in the 1960s a more permanent structure constructed using piers of the old railway bridge. The bridge is still open to walkers and locals who pass between Walberswick and Southwold.

In the 20th century sites adjacent to many of the county's bridges became gauging stations where a stream gauge is installed to monitor and test river water and to take hydrometric measurements of water levels. The Environment Agency regularly monitor the stations and keep them clear of vegetation and algae; even a small amount of either would affect the flow data.

At **Benhall** there is a river gauging station over the River Fromus and along the River Stour there are several including those at **Kedington, Glemsford** (West Hill), **Lamarsh** and **Langham**.

Glemsford is one of those reputedly haunted bridges. The ghostly figure of a monk is said to walk the footpath from the mill to the bridge but why, and whether or not it had anything to do with the nearby Monks Hall, has so far eluded local historians. It could also be a Victorian story arising from the discovery of skeletons at Scotchford Bridge in Glemsford during the 1850s. An ancient monastery was said to stand on land close to the bridge and a correspondent to the *Bury and Norwich Post* in January 1851 wrote:

> The skeletons were a male and young female, they ranged side by side, the male on the right side with no vestige of a coffin ... from their position east to west it implies it was a Christian burial which is confirmed by two sticks laid across them. It might have been a "strangers' corner" on a former burial ground.

There was also thought to be a Holy Well nearby and 13th century coins were found that were minted in the reign of Henry III. However, whether or not Scotchford is the same as the 'county bridge', also mentioned, is uncertain.

Beside the two-arch brick bridge at **Billingford** on the Suffolk-Norfolk border stands a large gauging station that monitors the River Waveney. Not far away, on the Norfolk side, is the old corn windmill: the little bridge was adequate for farmers with horses and carts, and even early tractors and trailers, but is unsuited to modern traffic. Billingford is one of several Suffolk boundary bridges to bear a 'cut mark', a broad, chiseled arrow below a horizontal line. Sometimes called bench marks, they were made by Ordnance Survey levelers in the 1800s to determine altitude above sea level. The arrowhead itself is the one used from the Middle Ages onwards to mark the property of the sovereign.

One of the best known, and perhaps most picturesque, gauging station and weir is at **Knettishall Heath** beside the single arch historic flint-faced and brick-patterned bridge over the River Little Ouse on the Suffolk-Norfolk border. The date of the bridge is unknown but it carries a number of worn shields and badges which would indicate that a nearby 18th century estate owner was responsible for its construction. As the main plaques differ on the two outer parapets it could also have been a joint project between the neighbouring counties although it would not have been cheap to build and is more likely to have been sponsored privately.

Knettishall Heath came into the ownership of the Suffolk Wildlife Trust in 2012.

Until relatively recently most river crossings were historic and entrenched in the traditional transport routes followed over hundreds of years to commercial or religious centres. Then technology enabled wider spans of both land and water to accommodate the increased number of vehicles and while the importance of some towns waned others grew. The navigations came and went; railway bridges proliferated and then declined; and heavy traffic became heavier, with

Heavy Goods Vehicles (HGVs) putting intolerable weights onto bridges built for horses and carts. Satellite navigation, when it became widely used, was and remains a menace to small bridges and quiet rural byways. The HGVs have become larger but the bridges become increasingly inadequate to the task.

The little bridge at **Coddenham** on the B1078 is hourly in danger. How it is still standing is a mystery as tractors and lorries thunder over it only missing its brick parapet edges by millimetres, edges that were built for another century entirely. Stout wooden posts have been placed either end of the beautifully curved brick construction but they are battered out of position and lean precariously to one side or the other. It is known as the Three Cocked Hat Bridge, or Three Cornered Hat, relating to bearings taken of three points of the compass which, if they do not quite meet up, form a triangle known as a 'cocked hat'. Urgent repairs were carried out on the bridge in 2013 and for several days the detours brought home the convenience of the bridge.

The bridge at **Shelley** on the west bank of the River Brett has protective bollards on both sides and flow detectors installed upstream. It was built in 1938 by the old East Suffolk County Council and its diminutive appearance belies the two massive steel girders below road level. Otters are often seen below the bridge.

Goldbrook Bridge at **Hoxne** has two speed reducing posts and a wide pavement on one side, in an effort to create a single track thereby reducing the wear and tear on the wooden fabric of the bridge. This is also another of the county's bridges that bears a cut mark, or Ordnance Survey Bench Mark (OSBM), on one of the pillars on the north side.

At **Great Wratting** the bridge has pilings 15½ metres deep on either side of the river to enable it to withstand the vibration caused by HGVs. In 2013 it was decided that the bridge was so badly abused by heavy traffic that it had to be demolished and rebuilt. The planners claimed that the three-tonne weight restriction on the bridge was frequently ignored and had made it unsafe. Work was complicated by the otters that inhabit that part of the River Stour. Crayfish reputedly breed prolifically at that point in the river but are a non native species which at some time escaped from farms near the mouth of the Stour.

There is still a ford that runs parallel with the bridge. It was probably there in Roman times that the Icenian queen Boudicca reputedly engaged and defeated the IXth Legion.

Close by, at **Great Thurlow**, a similar bridge and ford mirrors that at Great Wratting. This cast iron bridge was built by R Garrett & Son at their Leiston Works in 1851, the same year that Garretts had a prominent and well-stocked stand at the Great Exhibition in London. By then the company was being run by Richard

Garrett III, grandson of the company's founder, the first Richard Garrett.

Garretts had been engaged in bridge work from at least 1848: two of their bridge beams from that date can be seen at the Long Shop Museum in Leiston. The beam that once spanned the River Yox on the A12 (at Yoxford) was cast at Leiston and in 1983 the bridge was rebuilt by Jack Breheny Limited of Needham Market. The beam was noted by James Rickard of Yoxford who reported it to the Museum, whereupon Messrs Breheny presented it to the Long Shop.

The 1848 cast iron beam from East Bridge was recovered from East Bridge and presented to the Long Shop by A J Mew (Saxmundham) Metal Merchants on 17 September 1987.

The graceful red brick bridge parapets at **Chelsworth** have stout wooden posts at the entrance to the bridge and a three-tonne weight limit, but the bricks are regularly scoured and chunks gouged out of the wood. Although it is constantly being maintained it is considered 'at risk' of fatal injury. Not that this is anything new: on 8 January 1881 it was reported in the *Ipswich Journal* that Mr William Spooner, broker, and Mr William Barr, Innkeeper, both of Hadleigh were returning home from a sale at Chelsworth Rectory. It was a very dark night and the near wheel of their cart caught one of the posts on Chelsworth Bridge whereupon it overturned, pitching the two gentlemen out of the cart and onto the roadway. Mr Spooner sustained cuts to his head and was for some time 'insensible' while Mr Barr sustained injury to his left arm and a dislocated wrist. Mr Spooner revived sufficiently to drive the cart home as, happily, the horse was unhurt.

Before it was by-passed the small red brick single-arched bridge on the B1116 going out of **Framlingham** towards Wickham Market still bears the scars on the capping where it has been chipped and there are deep cracks in the walls.

The brickwork on the single arch bridge leading out of **Bungay** to Earsham (on the Norfolk side of the River Waveney) has scrapes and bashes along its inner parapets.

The early 18th century bridge at **Rushford** over the Little Ouse has sturdy protection posts in place after repairs were carried out in 2014. This single pointed arch bridge was described by Jervoise as both beautiful and picturesque and crosses the river on the boundary between Suffolk and Norfolk. It is Grade II listed but in 2014 it paid the price of belonging to an earlier age when it sustained damage to spandrels and the parapet wall on one side. Vital repairs were carried out to keep it safe although work was put on hold during May as birds were found to be nesting in a part of the structure which was to be demolished and rebuilt. The Wildlife and Countryside Act of 1981 decrees that any occupied bird nests cannot be destroyed until the chicks have fledged.

Rushford Bridge is thought to date from the late 18th century but was rebuilt in 1850 by the Shadwell Court Estate owned by the Buxton family. It was probably designed by the eminent Victorian architect, Samuel Sanders Teulon, who carried out extensions and remodelling at Shadwell Park between 1856 and 1860. The bridge is believed to have been built beside the existing ford from which Rushford takes its name. The parapet wall has two curious plaques in the shape of a triangle, or possibly a shape similar to a playing card spade, surrounded by a brick circle. This may be the mark of Samuel Teulon or perhaps a 'seal ring' relating to the Buxtons.

In the park at Shadwell (Norfolk) is the Holy Well of St Chad, which would have attracted medieval pilgrims on their to or from Our Lady at Walsingham in Norfolk. There was unlikely to have been a bridge then so pilgrims would have used the ford.

The little brick Pentlow Bridge at **Cavendish** on the Suffolk-Essex border has to cope with an almost constant stream of day-time traffic. It has the date 1886 on a very worn plaque on the outer edge of the parapet. On the Essex side there is a long raised walkway, indicating that the Stour is likely to flood at that point.

It is the practice of Suffolk County Council to give bridges a general inspection every two years and a more detailed inspection every six to ten years. Until 1974 the east and west of the county were administered by their own councils which often led to joint projects not only between the two Suffolk councils but also with neighbouring counties.

A bridge near Newmarket at **Sipsey** crosses the River Stour on the B1061 and is marked on Hodskinson's 1783 map. The road forms the border between Suffolk and Cambridgeshire and the bridge was a joint effort between West Suffolk County Council and Cambridgeshire County Council in 1923. There is a plaque to each county on the parapets.

The original name of the bridge is lost in the mists of time but in 1954 a racehorse named 'Sipsey Bridge' was born at a Yorkshire Stud. It went on to have an illustrious career and initiated a long line still traceable today.

In 2003 a steel truss bridge was erected at **Mendham** over the River Waveney to replace the old steel riveted bridge that was built in 1914. Traffic had gradually weakened the old bridge and from 1999 was subject to a 3 ton weight restriction although that was not enough to prevent wear and tear to the fabric which rendered it potentially unsafe.

The 1914 bridge was scheduled for closure in July 2003 but the project was delayed by almost a month when swallows were found to be nesting underneath. However, in August the old bridge was closed and worked started on the new one. The 1914 commemorative stones, showing the castle emblem of the old East Suffolk County Council and the county badge of the Norfolk County Council, were built into the end pillars of the new bridge.

It is to be wondered what the artist Alfred James Munnings (later Sir) would think of the dramatic modern steel structure of the bridge with its beams and struts rising up over the river. Munnings was born in 1878 at Mendham where his father ran a water mill on the river. There is very little bridge history associated with Mendham but it may be assumed that the mill at least would have had a bridge of some kind and it is likely that the nearby Cluniac Priory, founded in 1140, would have bridged the river early in its foundation. One of the Mendham bridges may well have inspired Munnings towards his artistic career and as a painter of rural scenes he often painted bridges.

One of his more famous paintings is *Sketching at Wiston Bridge* and shows his friend and fellow artist Maurice Codner at work with his easel beside a wooden cattle bridge surrounded by willow trees. (Wiston was previous known as **Wissington** and is now united as a single parish with neighbouring Nayland.) There were three such paintings and one was exhibited at the Royal Academy in 1946.

During the 20th century many bridges were widened or reconstructed to cope with increased traffic. The modified bridge at **Brandon**, for example, was opened on 23 July 1954 by the Minister of Transport, the Rt Hon Alan T Lennox-Boyd, MP, accompanied by various members of the Highways and Bridges Committee. Happily the new bridge retained pedestrian refuges, two on each side, which had been such a feature of the old, picturesque bridge that it replaced. Brandon is one of only a handful of Suffolk bridges that have pedestrian refuges in their parapets.

Brandon was one of the bridges included in the Survey of Ancient Bridges that was carried out over many years by Edwyn Jervoise under the auspices of the Society for the Protection of Ancient Buildings. He visited over 5,000 bridges in England and Wales and his findings were published in a series of books in the 1930s.

In December 1984 Suffolk County Council agreed how best to enact the European Union (EU) legislation requiring all bridges to be assessed to check whether they had the carrying capacity required by 40 tonne lorries. Upgrades to local bridges were not popular and still cause controversy when weight restrictions are imposed or cherished structures altered. As recently as February

2016 almost 3,000 people signed a petition aimed at halting planned works to the Silver Street Bridge at **Kedington**.

In 1999 the bridge at **Kentford** was one of those upgraded by Suffolk County Council to comply with the EU's 40 tonne weight regulation. The village sign, standing to the west of the bridge and on the county border, is unusual in that it is double-sided and shows Kentford on the Suffolk side and Kennet on the Cambridgeshire side. The piers of an older structure can still be seen poking out into the river, possibly that built in 1948. It is somewhat confusing, though, as in 1929 the County Surveyor changed the place where steam lorries drew water up from the river and the masonry could date from that time.

The bridge over the River Alde on the B1119 linking the villages of **Rendham** and **Sweffling** was upgraded in the face of local opposition in 2003. Built in 1927 it failed a structural inspection in 1998. It had been restricted to traffic under 3 tonnes since 1999 and was rebuilt over a 5 month period at a cost of £375,000. Although the route was considered unsuitable for heavy lorries it also meant that fire engines, school buses, farm tractors, road gritters and ambulances had been forced to use other roads through neighbouring parishes.

Among the spoil generated by the works was found a 17th century canon ball near to the east side of the bridge on the north bank of the river. The director of Suffolk Underwater Studies, Stuart Bacon, identified it as being made of iron, possibly with a heavier metal such as lead inside, and probably a naval cannon ball. However, its weight indicated that it had not been immersed in water for any length of time since water leaches the metals eventually rendering the material practically weightless.

It might have come up in a fishing net around the time of the Battle of Sole Bay at Southwold in 1672 when English and French fleets fought the Dutch in a fierce action. It was almost certainly fired from a ship close to land as cannon balls used by one ship against another could afford to be heavier than those for long range use or in the open sea.

Theories about how the cannon ball came to be buried beside Rendham bridge are several but in all likelihood it was brought to Rendham in a consignment of rubble or shingle from somewhere on the coast and used for repairs to an earlier bridge.

Rendham bridge is one of those depicted on the village sign.

A more unusual bridge reference is found on the village sign at **Rattlesden** which depicts a pair of 19th century whale bones. These represent two whale bones that were placed at the bridge in the early 1800s by the owners of a nearby house. They stood over this part of the River Rat, a tributary to the River Gipping, until 2000 when they were replaced by wooden replicas.

In the Victorian era there was a craze for erecting whale bones across the country but there is only one pair of originals left in Suffolk, that the Red Lion pub in **Great Wratting**. It is possible those at Rattlesden were somehow connected with the short-lived whaling operation at Ipswich where whales from Greenland were processed on the banks of the Orwell. In the 1780s two ships, the *Ipswich* and the *Orwell* began whaling and on her first trip the *Orwell* caught seven whales, which netted 4 cwt of whalebone and 150 butts of blubber. Subsequent ventures were unsuccessful and the operation folded soon afterwards.

The long and varied history of tolls on Suffolk bridges drew to a close in 1962 when on 16th March, Hansard recorded a debate on the subject:

> While [this House] recognises that there may be new circumstances in which it is sometimes desirable to charge tolls on some bridges, [it] urges the Government and the highway authorities to take the earliest practicable steps to extinguish those tolls which have existed for many years on certain bridges and which now have no valid purpose.

At this time there were only 31 toll bridges still in operation dotted around England, Wales and Scotland, the one at **Syleham** on the Suffolk-Norfolk border being one of them and the only one in Suffolk. Across the country tolls on many bridges had been abolished by having been bought by the local County Council who then declared them 'county bridges'. The Syleham-Brockdish (in Norfolk) toll was still operating in the 1950s when the toll was 1d (one old penny) but was waived for those on foot or with bicycles who worked at the nearby Mill. There was a Toll Cottage beside the bridge until 1928 when it, and much of the nearby Mill and drabbett factory, was destroyed by fire.

During the parliamentary debate it was stated that the Syleham toll had been in place since 1727 although that is likely to have been the date it was documented rather than when it was first instituted. The practice of charging a toll on that crossing would have gone much further back probably to pre-Reformation times when Thetford Priory held a manor in Syleham, granted to them by Herbert de Losinga, first Bishop of Norwich. An earlier version of the present Monks Hall was a hunting lodge for the Cluniac monks and it is likely that they had a presence at St Mary's Church (variously called St Margaret's). In the 17th century there was a wooden bridge across the river to the north-west of the church, connecting it to a causeway on the Norfolk side and probably originally for the use of the monks. There is no sign of it now, the only crossing being the one near the old mill. Religious establishments were invariably the first to establish a toll especially in places where river crossings were few and far between. Since the toll house was on the south side of the river, at Syleham rather than Brockdish (Norfolk), it would seem that the monks instigated the toll to be paid by the secular users while visitors to the church or Hall came via the monks' bridge.

A bridge over the Waveney would certainly have been required in the 12th century when in 1174 Henry II encamped at Syleham with 500 carpenters to prepare siege engines for the attack on Hugh Bigod's Bungay and Framlingham castles. Bigod, 2nd Earl of Norfolk, is said to have sworn loyalty to the king and surrendered his castles at a meeting in Syleham.

The existing weir is called Toll Bridge Weir but there is now no trace of anything to do with the toll other than the small brick bridge, its capping stones covered in green moss.

Jervoise recorded the date of Syleham bridge as 1847 but thought it 'uninteresting' though it is hardly that. What then appeared to be a mundane, work-a-day bridge is now considered a picturesque feature in a modern landscape.

Although there are a few private bridge tolls still in existence in England, Syleham is not one of them, it having been retrieved by East Suffolk County Council, without compensation to the then owners, in 1959.

There is another curious toll that survived until 1907. In that year a surveyor, H Miller, made a note that a chain and padlock were fixed in the centre of **Homersfield** bridge for one day a year, generally when the river was in full flood. A toll of two pence was charged although foot passengers were allowed to pass freely. It is likely that this single day derived from Sir Shafto's year-round right of toll initiated in the 1760s when the bridge was built. Some time after 1907 liability for the bridge lapsed: the structure fell into serious disrepair and the annual toll tradition ceased

Thelnetham bridge on the southern bank of the Little Ouse

Mendham bridge over the River Waveney

Syleham Mill bridge at the site of the old toll road

Work being carried out on one of the eastern piers of the Orwell Bridge

Approaching the Orwell Bridge

The UKD dredger 'Orca' passing through the raised Lowestoft bridge

The lifting bridge at Lowestoft brings town traffic to a halt

The Writer's Bridge at Henham Park

Repairs being carried out at Mells bridge in 2016

*In 2016 a mud bank was removed from the southern aspect of Abbots Bridge,
Bury St Edmunds*

Hemingstone bridge with an 1813 keystone

Needham Market's Gipping and Hawks Mill
bridges leading to the 1922 navigation bridge

Hump back bridge at Baylham Mill

Sign beside the ford and bridge at Great Wratting,
thought to be the only one of its kind in Suffolk

CHAPTER 4

In 1982 Suffolk was once more in the forefront of bridge technology when the **Orwell Bridge** was opened to traffic. It is in fact two parallel bridges, carried on 19 sets of piers, and crosses the River Orwell between New Channel and Freston Reach. The bridges have a gap of just a few inches between them. At the time of construction its 190 metre span was the longest pre-stressed concrete structure in the world. It is built of concrete box girders that allow movement for expansion and contraction and took two years to build. Pilings were sunk 40 metres into the river bottom. The main contractor was Stevin Construction, a Dutch company, which began work in October 1979. The project was funded by the Department of Transport and cost £24 million.

Car drivers regularly complain that the solid bridge parapets obstruct their view of the glorious vista of the River Orwell. Only drivers of HGVs, vans and some four wheel drive vehicles can see anything at all. It was thought for some time to be a safety feature in that drivers would keep their eyes on the road rather than the view but in 2016 it came to light that during construction a Stevins engineer explained that the parapets had to be high enough to allow for shipping to pass underneath and for two years the company had monitored the wind tunnel caused by the river. Because the A14 bridge had to carry container traffic to and from Felixstowe Docks the height of the concrete parapets were crucial in preventing high-sided vehicles being blown over by high winds.

The Orwell Bridge was built to the highest specifications of its time and its importance to the county is only realised when an accident causes it to be closed or there is road or bridge maintenance being carried out. It stretches dramatically from bank to bank and, discounting the legendary attempts by the Romans to ford the river, is now the first crossing of the Orwell.

A down side of the Orwell Bridge is the high suicide rate associated with it. More than 40 people have died in the waters below the bridge since it was opened in 1982.

When the new 3-span Ballingdon Bridge in **Sudbury** was completed in 2003 it was the first trunk road bridge to be built in Britain with an architect leading the design team. It is Suffolk's answer to the 'mega-bridge' with a construction that uses aluminium in sections of the balustrades. These are capable of stopping a 42-tonne truck yet have the appearance of a simple hand rail.

This is the latest in generations of wooden and concrete bridges that have spanned the River Stour at this point, where there has been a crossing place for centuries. This one was built to last 120 years. The design, by Brookes Stacey Randall architects and Ove Arup civil engineers, was chosen by the people of Sudbury as the result of a competition. Use of this busy bridge was maintained throughout the construction period, with each complete side being erected in a single weekend.

The bridge has special bricks underneath it which provide homes for bats and although the special lighting is part of the design, it does claim to have minimum light pollution by reducing reflections on the water.

A drawing of the 1805 Ballingdon Bridge appeared in the Proceedings of the Suffolk Institute of Archaeology and History in 1892. This shows it to have been a simple wooden structure with six piles and raised in the centre. It was the last timber bridge and was dismantled in 1911 to be replaced by a reinforced concrete structure, known now as the Old Ballingdon Bridge.

Unusually, the previous generations of timber bridges followed an 8-arched brick structure built in 1521 that was swept away by floods in 1594. Perhaps it was thought that the periodic replacement of a wooden bridge was cheaper than the continued maintenance of a brick one.

The Lemons Hill Bridge at **Tattingstone** crosses Alton Water, a manmade reservoir with a circumference of over eight miles that is the largest area of inland water in Suffolk. A water shortage in the 1960s led to twenty potential water storage sites being surveyed and Alton was finally chosen. The village was split in half when the valley was flooded in the 1970s. Alton Hall, twenty houses and two farms were inundated but Alton Mill was dismantled and rebuilt at the Museum of East Anglian Life at Stowmarket. The reservoir took thirteen years to construct and fill with water.

Tattingstone Heath is connected to what is left of the parish by the Lemons Hill Bridge which has the distinction of forming a crossing over an entirely new body of water. The two-lane highway is not absolutely straight, its gentle curve picked out by white metal railings.

Broadly speaking there are five main ways to construct a bridge: by arches, girders, cantilever, Bascule or suspension. The pedestrian Sir Bobby Robson Bridge in **Ipswich** is cable-stayed and its 200ft (60 metre) height makes it the second highest structure in Ipswich. It hangs from a 35 metre high 'wishbone' pylon with eight suspension wires in four parallel pairs. The bridge was built in 2009 at a cost of £800,000 and named after Sir Bobby Robson, the ex-Ipswich Town Football Club manager and England's national football team manager. The name was chosen following the *Evening Star* newspaper's competition.

This is a traffic-less footbridge where you can linger without fear of your life and watch the water tumble over the old weir at low tide on one side and catch a glimpse of the disused railway bridge on the other. The soffit, or underside, of the bridge is designed to allow for clearance of debris during possible flooding. During the last few years Little Egrets, which have only been breeding in this country since the 1990s, have become a familiar sight feeding on the mud at low tide both here and close to the Princes Street bridge.

In June 2016 it was the venue for the Suffolk Soul Singers when they took part in the BBC's Music Day 'Take it to the Bridge', one of 40 such events taking place on 40 bridges across Britain.

Apart from the Orwell Bridge, the Mutford crossing at **Oulton Broad** and the lifting bridge at **Lowestoft** are probably the busiest in Suffolk. Both are unusual for Suffolk in being bascule, or moving, road bridges (the swing bridge at Somerleyton is a railway bridge built in 1905).

The present Mutford Bridge was opened on 17 November 1992, the Chairman of Suffolk County Council Highways Committee, Mr G K D McGregor, officiating. It is the latest in a series of crossings at that point.

The first brick bridge, replacing a wooden one, was built in 1760 and fishing and small trading vessels had to lower their masts to get underneath its span. It was swept away in a tidal flood in 1791 and for a year a ferry was the only method of crossing Lake Lothing. The next to be built was in 1827 when the modern shape of Oulton Broad was formed with the construction of the lock. The dyke between Oulton Broad and the Waveney was deepened and a canal was cut between the rivers Yare and Waveney. In 1830 the first moveable bridge was constructed followed in 1939 by a swing bridge which was the first all-welded bridge in the country and cost £45,000. It was demolished to make way for the bascule bridge.

There is a history of the bridge on a nearby bollard. It is quite badly corroded, though just readable. Several bridges, including the railway bridge, can be seen from the various walkways and banks in the immediate area as well as boats of all descriptions on either side.

Tons of spoil from the new lake being dug on the Stradbroke Estate at **Henham** was used in the construction of the 1992 Mutford Bridge and the lake itself got its own bridge in 2011. It is constructed primarily of wood and steel, its wavy design intended to reflect the lake. Named the Writer's Bridge it replaced a pontoon that was erected specifically for the duration of the annual Latitude Festival held on the Estate. In 2010 it was said that at times just the amount of people on the floating bridge at one time meant it was under pressure and in danger of submerging it.

Lowestoft had no bridge until 1830 when the Norwich & Lowestoft Navigation Company cut a channel from Lake Lothing to the North Sea. The resulting harbour was made up of two sections: the Inner Harbour formed by the originally land-locked Lake Lothing, and the Outer Harbour that was formed by the breakwaters. The company also constructed a cast iron bridge, that opened in the centre, to link the north and south areas of the town. This bridge, which lasted for 67 years, was shipped round from Great Yarmouth (Norfolk) by river during which journey it became trapped for a while in ice between Somerleyton and Oulton Broad.

The 1830 bridge was replaced in 1897 to celebrate the Diamond Jubilee of Queen Victoria. The single span swing bridge was constructed by the Great Eastern Railway Company and although it was officially called the Victoria Bridge, the name never took off and it remained known as just The Bridge.

The third bridge was opened in 1972 by the local Member of Parliament, the late Rt Hon Jim Prior (later Baron). By 1968 the Victorian bridge was past its best and the town began lobbying for a new bridge. However, it took so long to get the bridge financed and built that in January 1969 a 60-ton Bailey bridge had to be erected by the Royal Engineers so that people could get from one side of the town to the other. Fifty men of the 20th Field Squadron worked non-stop for two days to set up the bridge across the Inner Harbour channel. Originally it had been estimated that it would take a week to build but in the event it took the engineers just 48 hours. The Mayor of Lowestoft and Jim Prior sent the men a consignment of beer and the National Mission to Deep Sea Fishermen provided a supply of tea.

Jim Prior had fought hard to get the new bridge and in a Commons Debate on 18th July 1963 said:

> The swing bridge carries the traffic on the main A12 London to Great Yarmouth trunk road, across Lowestoft harbour, and it is also a vital link to local traffic, including large numbers of industrial workers who live on one side of the bridge and work on the other.

At the time the state of the bridge was so bad that one of Mr Prior's constituents had written to the local paper pointing out that in the last 17 years he had spent all the working hours of one year waiting for the bridge to close so that he could get across. In 1961 Mr Prior had helped to prevent a strike resulting from the frustration caused by delays on the bridge:

> So great are the delays in the summer that car drivers at the end of the

queue waiting to cross the bridge find that it has opened again before they have reached it. The result is that it can take an hour or one-and-a-half hours to get across the bridge.

Motorists today still complain about the amount of time wasted when the bridge is lifted to allow shipping to pass, which it does several times a day. Relief for Lowestoft's traffic problems was promised in the 2016 Budget when it was announced that the go-ahead was given for a new crossing over the harbour entrance. This will hopefully alleviate the traffic problems in the town exacerbated by the bascule bridge.

Another Lowestoft bridge is the Ravine footbridge that was presented to the town by the first official Mayor of Lowestoft, Alderman William Youngman, to commemorate Queen Victorian's Golden Jubilee in 1887. It links North Parade and Belle Vue Park. Originally the park was created from a piece of 'bleach' land, an area where the public could dry their washing. A fine picture of the Ravine Bridge by the 19th century Lowestoft-born artist, George Vemply Burwood, hangs in Lowestoft and East Suffolk Maritime Museum. The artist was chosen by the town to give the loyal address to the Queen on the occasion of her Jubilee.

There may also have been a bridge at Kirkley which was used by the late 18th century coaches from London. This followed an old circuitous coaching route to Pakefield via Mutford Bridge. Later, private coaches by-passed Mutford Bridge and, the story goes, the new route went along the foreshore path by the beach crossing east of Lake Lothing. Christopher J Brooks wrote:

> Sometimes a man stood there with boots on to carry pedestrian travellers to Lowestoft and Pakefield fairs over a small outlet of Lake Lothing running to the sea ... a crossing by the lake was to be canalised and bear a bridge over it by 1831.

It stands to reason that if a parish is proud of its bridge it will be included on the village sign. That at **Debenham** was erected in 1977 in celebration of the Silver Jubilee of Her Majesty Queen Elizabeth II. The **Great Bealings** bridge over the River Lark appears on the village sign and that at **Ixworth** depicts a man leading three packhorses to denote the drovers' routes of the 1800s.

Given its very long history as a crossing point, it is not surprising that **Snape** has a bridge on its sign. The bridge also carries a green mushroom-headed OSBM 'bolt', another version of the cut or bench marks.

The sign at **Dalham** shows a white railed footbridge which represents a series of such bridges crossing the River Kennet, which rises near the village and runs through its centre. The bridge beside the Affleck Arms is a road bridge but there are several footbridges running off both sides of the road, the village being a good example of a Suffolk linear settlement.

White bridge railings abound all over the county. Sometimes they herald a modern iron girder underneath but sometimes there lies hidden a charming legacy from a bygone age. The two arch bridge at **Kettlebaston** for example appears to be just another white railed bridge, but closer examination reveals the rounded tops of the older parapets now at road level which at one time stood proud of the crossing.

White railings run off the ends of brick parapets at the old mill in **Needham Market** which has a two arch bridge beneath; at **Thelnethham** the road bridge and the raised Irish bridge are both picked out by white rails.

The bridge at **Higham**, over the River Brett, was built in 1846 and has very stylist rails between its brick pillars and extended parapets.

There are three small bridges around **Wainford** in the Waveney Valley on the Suffolk-Norfolk border which are edged by white railings. The single-arch boundary bridge at Grain Mill is hump backed with a high peak. They were all were built at a time when farmers transported grain to the mills by horse and cart.

A similar arrangement is seen at **Laxfield** where several white-railed footbridges criss-cross the village, together with a single road bridge.

The Thurleston Lane bridge at **Whitton** (Ipswich) has only a few railings and its modern repairs of the parapet capping at road level disguise a charming little single-arched bridge.

Walsham-le-Willows has several bridges as the stream flows through the village centre and those in the main street are distinguished by white railings. At Miller's Bridge the railings extend either side of the bridge, which was reconstructed in 1938 by what was then the West Suffolk County Council. The surveyor was E H Bond, who also appears on the plaque on the bridge at Wormingford. There is a local dispute about whether the bridge was named for a miller or for a family called Millar: the bridge plaque is spelled with an 'e' but the nearby road with an 'a'.

On 21st March 2012 a white willow was planted beside Miller's Bridge to celebrate the Diamond Jubilee of Queen Elizabeth II.

Just outside the village Old Brook Bridge stands next to the new one which is marked on the road by unpainted railings. The old bridge is also near a footbridge and an Irish Bridge and is one of those where you can stand and stare without risk to life and limb, although the volume and noise of passing traffic is not particularly conducive to contemplation.

Playford bridge has white railings atop a modern concrete and girder construction over the River Fynn but a packhorse bridge once reputedly stood here, long since demolished.

The bridge that straddles the River Box at **Boxford** was reconstructed in 2009 and retains its white railings as does the Waldingfield bridge, just west of **Little Waldingfield.** The latter was constructed in 1936 under the direction of the county surveyor, Edward Herring 'E H' Bond, based at Bury St Edmunds.

The Boxford Bridge sports an oval plaque on part of the old roadway but the inscription is unreadable as it is, unfortunately, sited immediately below a drain outlet pipe which has eroded and blackened it irrevocably.

The county surveyor, E H Bond, was also responsible for the bridge at **Great Bradley**, the first village in Suffolk on the River Stour after leaving its source in Cambridgeshire. The bridge was reconstructed in 1935.

Not all small country bridges have railings, of course, and the bridge at **Bildeston** is hardly noticeable, its parapets having been incorporated into the lines of fencing and walls either side of the road. It must surely hold the record for the highest parapets, however, which stand some six feet high.

At the top end of a narrow road leading from **Hoxne** street, opposite the school, up to the Stradbroke road is an endearing double arched brick crossing which for its size has good solid parapets.

The bridge at **Hemingstone** has a mix of wood and metal railings. A weathered key stone is incorporated into the modern brickwork which bears the date 1813. In May 2013 the villagers gave their bridge a 200th birthday party to celebrate the its reopening after repairs which had recently been undertaken.

Surveyors from Suffolk County Council's Highways Department visit each bridge every two years during which its condition is recorded. Every six to ten year more detailed inspections are undertaken, thus the state of the county's bridges are monitored and at any given time several are undergoing repair, rebuilding or upgrading. Bridges, of course, are as susceptible as they ever were to damage from flooding but the primary cause of damage is down to traffic. In January 2017 repairs were under way on the Boot Street bridge at **Great Bealings** to rebuild the parapets of the little red bricked hump back bridge that had succumbed to traffic damage of a kind that did not exist in the 1840s when it was first built. The plan is to give priority to westbound vehicles once the work is completed.

On-going renovations are also being carried out at Hempyard bridge at **Ixworth,** where access is limited to all traffic except pedestrians, and sections of the bridge at **Parham** were found unsafe at the end of 2016.

The approaches to one of the county's most evocative bridges, the 12th century Abbots Bridge in **Bury St Edmunds**, had a face lift in 2016 when a mud bank was removed by a specialist contractor. It had built up over some twenty years obscuring the view of the south aspect. That side of the bridge forms part of the Abbey Gardens which are the remnants of the Benedictine Monastery that was despoiled at the Dissolution.

Bridge history is, of course, still evolving, as witness plans for the new river crossings in **Ipswich**, currently under discussion. The proposal is for a new bascule bridge from the West Bank to Cliff Quay. Another smaller bridge is planned over the New Cut between Felaw Street and the island site, and a new foot and cycle bridge will span the Wet Lock gates. Although most of the money will come from central government, Suffolk County Council will still be asked to approve a considerable contribution with additional amounts from the New Anglia Local Enterprise Partnership and businesses in and around the town. It is hoped that the project will come to fruition in 2020.

Meanwhile, the Orwell Bridge continues to set new records. In 2016 one of the largest ships to call at the Port of Ipswich, the 515ft-long *Dijksgracht*, entered the estuary carrying a shipment of 10,500 tonnes of rice from Texas bound for the Ipswich Grain Terminal. On what was a low spring tide its progress had to coincide with a 25 minute window of safe passage and even then everyone held their breath, as even one small wave could have been disastrous. The vessel cleared the bridge by just fourteen inches.

BIBLIOGRAPHY and INTERNET SOURCES

Bench Mark Database *www.bench-marks.org.uk*

British Listed Buildings *www.britishlistedbuildings.co.uk*

Brooks, Christopher J *China Returns to Lowestoft* (2001)

Cook, Olive *Suffolk* (1948)

Cooper, Alan *Bridges, Law and Power in Medieval England, 700-1400* (2006)

Cruickshank, Dan *Bridges: Heroic Designs that Changed the World* (2010)

Defoe, Daniel *Tour Through the Eastern Counties* (Originally published 1724)

Diary of John Longe, Vicar of Coddenham, 1765-1834 (Edited Michael Stone, 2008)

Hodskinson, Joseph *Hodskinson's Map of Suffolk in 1783* (2003 edition)

Jervoise, E *The Ancient Bridges of Mid and Eastern England* (1932)

Kirby, John *The Suffolk Traveller, 1735* (Suffolk Records Society 2004)

Proceedings of the Suffolk Institute of Archaeology and History (various)

Reyce, Robert *Suffolk in the XVII Century* (1618)

Scarfe, Norman *The Suffolk Landscape* (2002)

Suckling, Alfred *The History and Antiquities of the County of Suffolk Volumes 1 and 2* (1846)

Taylor, Dr J E *In and About Ancient Ipswich* (1888)

Wallace, Doreen *East Anglia* (1939)

Weston, D E *Lord Bristol's Amazing Steam Venture as seen by Mr William Howlett from 1889 to 1915* (1981)

Wodderspoon, John *Memorials of the Ancient Town of Ipswich* (1850)

How to find Suffolk Towns and Villages Suffolk Family History Society (2000)

The Suffolk Village Book Suffolk Federation of Women's Institutes (1991)

Whitton's Thurleston Lane bridge,
typical of many small bridges across the county